AF699241

WITHDRAWN

CONVECTION IN ROTATING FLUIDS

FLUID MECHANICS AND ITS APPLICATIONS
Volume 29

Series Editor: R. MOREAU
MADYLAM
Ecole Nationale Supérieure d'Hydraulique de Grenoble
Boîte Postale 95
38402 Saint Martin d'Hères Cedex, France

Aims and Scope of the Series

The purpose of this series is to focus on subjects in which fluid mechanics plays a fundamental role.

As well as the more traditional applications of aeronautics, hydraulics, heat and mass transfer etc., books will be published dealing with topics which are currently in a state of rapid development, such as turbulence, suspensions and multiphase fluids, super and hypersonic flows and numerical modelling techniques.

It is a widely held view that it is the interdisciplinary subjects that will receive intense scientific attention, bringing them to the forefront of technological advancement. Fluids have the ability to transport matter and its properties as well as transmit force, therefore fluid mechanics is a subject that is particulary open to cross fertilisation with other sciences and disciplines of engineering. The subject of fluid mechanics will be highly relevant in domains such as chemical, metallurgical, biological and ecological engineering. This series is particularly open to such new multidisciplinary domains.

The median level of presentation is the first year graduate student. Some texts are monographs defining the current state of a field; others are accessible to final year undergraduates; but essentially the emphasis is on readability and clarity.

For a list of related mechanics titles, see final pages.

Convection in Rotating Fluids

by

B. M. BOUBNOV and G. S. GOLITSYN

Institute of Atmospheric Physics,
Russian Academy of Sciences,
Moscow, Russia

KLUWER ACADEMIC PUBLISHERS

DORDRECHT / BOSTON / LONDON

A C.I.P. Catalogue record for this book is available from the Library of Congress.

QC
151
.B58
1995

ISBN 0-7923-3371-3

Published by Kluwer Academic Publishers,
P.O. Box 17, 3300 AA Dordrecht, The Netherlands.

Kluwer Academic Publishers incorporates
the publishing programmes of
D. Reidel, Martinus Nijhoff, Dr W. Junk and MTP Press.

Sold and distributed in the U.S.A. and Canada
by Kluwer Academic Publishers,
101 Philip Drive, Norwell, MA 02061, U.S.A.

In all other countries, sold and distributed
by Kluwer Academic Publishers Group,
P.O. Box 322, 3300 AH Dordrecht, The Netherlands.

Printed on acid-free paper

All Rights Reserved
© 1995 Kluwer Academic Publishers
No part of the material protected by this copyright notice may be reproduced or
utilized in any form or by any means, electronic or mechanical,
including photocopying, recording or by any information storage and
retrieval system, without written permission from the copyright owner.

Printed in the Netherlands

Contents

INTRODUCTION

Spatial inhomogeneity of heating of fluids in the gravity field is the cause of all motions in nature: in the atmosphere and oceans on Earth, in astrophysical and planetary objects. All natural objects are rotating which must be taken into account at sufficiently large rotation rates. Therefore, the first question is when the rotation of convective systems is negligible and when it must be taken into account and to what changes it would lead in flow patterns and in the transport properties of the convective fluids.

For most technical applications and for the sufficiently small scale geophysical phenomena the Earth's rotation is not important. However, in some technical and technological applications one may encounter with sufficiently fast rotation and here many interesting phenomena may arise which are still explored only very poorly, but which may have potentially important applications.

For geophysical and astrophysical objects the rotation together with spatial inhomogeneity of energy sources distribution in the gravity field determines completely the origin and character of the fluid motion. It leads to the general circulation of the atmosphere and ocean, internal motions in the Earth's liquid core (for its mantle the role of rotation is negligible). Among astrophysical objects we may mention fast rotating pulsars, accretion discs, planetary atmospheres. The flow patterns in them manifested through some observable phenomena, the transport of heat, energy in general, and matter by such motions all are of great interest.

It so happened that the convection in rotating fluids up to quite recently was a neglected subject, especially the simplest case of a plane horizontal rotating layer. Here we hope to demonstrate that even this simplest case can reveal a rich wealth of the hydrodynamic regimes and phenomena: beautiful laminar coherent patterns and how they evolve into a turbulent regime with its own statistical regularities, vortical interactions and instabilities.

The second part of the book deals with the convection in a differentially heated rotating annulus which is used about forty years already to simulate the general circulation of the atmosphere. The subject is much more developed than the simple case of convection in the plane layer. But new recent developments here are of importance and interest.

The beginning of the history of studies of convective fluids in rotation can be traced to the famous work of 1732 by Hadley who tried to explain the origin of the trade winds. The early ideas on the origin and structure of the atmospheric general circulation can be found at Lorenz (1969).

The beginning of the systematic studies of convection in the plane horizontal layer without rotation started with experimental work by Boussinesq (1900) and theoretical paper by Lord Rayleigh (1916) who studied the origin of convection in a layer heated from below. Similar studies of convection in the layer with rotation along the vertical axis have been carried out by Chandrasekhar (1953, 1961) and independently by Nakagawa & Frenzen (1955). Chandrasekhar carried out systematic theoretical studies of the onset of convection for various boundary conditions: both horizontal boundaries are free or rigid, and the lower boundary rigid and the upper one free. The second paper considered only the simplest case of both boundaries free but they also conducted some experiments which partially confirmed the shape of the stability curve. Then for about forty years the experimental studies had been neglected until the present authors started a systematic program of experimental studies of the convection in a horizontal convective layer (see Boubnov & Golitsyn, 1985a, 1986, 1990a, 1991). This has been preceded by two theoretical works by Golitsyn (1980a, 1981) where the notion of geostrophic convection has been introduced, some self-similar regularities have been proposed for the velocity scale of convective motions which were confirmed by very crude "kitchen" experiments b(performed at his kitchen). These "kitchen" experiment findings later on were confirmed in a series of special laboratory experiments by Boubnov (1988, 1989), Boubnov and Golitsyn (1990b) and Fernando et al (1991).

That the studies of convective motions began with the case of the plane layer, one may easily understand considering mutual orientation of the main vectors determining the fluid motion. For the thermal convection the main directions are the gravity acceleration and the temperature gradient, i.e. the heat or buoyancy flux. The rotation adds the angular velocity vector and the simplest case is the horizontal fluid layer heated from below and rotating around a vertical axis. Then all three vectors are parallel to each other.

The second simple case is when the gravity acceleration and the angular rotation vector are parallel but the temperature gradient is perpendicular to them. The first experimental realizations of this case have been performed in the early 1950s by Fultz and Hide. These experiments were directed to modelling the atmospheric general circulation and convective motions in the Earth's liquid core. Soon afterwards these experiments initiated a very large number of theoretical and experimental studies and served as basic experiments in laboratory studies of the baroclinic instability. Most of the experiments were conducted in rotating annuli. Here, as in the plane horizontal rotating layer, one can observe a variety of flow regimes and the theoretical description is successful only in a limited number of regimes.

This book starts with general equations for studying convection in rotating fluids (Chapter 1). In Chapter 2 the consideration is limited to the Rayleigh numbers when the linear approximations is valid and the problem can be treated analytically up to (or almost up to) the end. The stability curves are found for all kinds of boundary conditions for velocity and temperature fields and the most unstable wave numbers are found. At large supercritical Rayleigh numbers the motion is irregular, not only analytical but also computational methods are practically impossible and the experimental studies lead as a rule the theoretical considerations. However the irregular motions reveal at the same time new statistical regularities established by dimensional analysis. The existence of self-similar regimes in convection of rotating fluids was first shown theoretically in 1980 and has been confirmed by experiment.

Chapter 3 deals with rotating annuli with fluid heated inhomogeneously . The classical case is of the horizontal mean temperature gradient, but we also consider (Chapter 4) the case when a vertical component of this gradient is present which has important geophysical implications (such as "nuclear winter" and global dust storms on Mars). An analysis of theoretical studies for rotating annulus experiments is carried out and as a base model for this a two-layer geostrophic model is taken, generalizing the one first developed by Lorenz (1960). This approach allows us to describe, at least qualitatively, all sorts of flow regimes observed in laboratory experiments.

Chapter 5 deals with localized buoyancy sources in rotating fluids which can generate intense singular vortices. The next two chapters consider centrifugal effects which prevail at large rotation rates over gravity and the spherical geometry. This last problem is considered in detail in the classical book by Chandrasekhar (1961) and in the review by Yavorskaya and Belyaev (1982). We only add some important material published during the last decade.

The final Chapter contains applications of the results to some geophysical and astrophysical objects.

Working since the end of the 1970s on the subject, the authors in their studies and in this book always tried to elucidate the physical essence of the phenomena, to determine and classify flow regimes in the space of similarity numbers. The theoretical and computational results are presented here in some detail only in those cases when the results help to explain basic qualitative motion characteristics, otherwise they are described only shortly.

Chapter 1
GENERAL NOTIONS

1.1 Description of the rotating fluid motion. The Boussinesq approximation

We consider the motion of a thermally inhomogeneous fluid rotating with a constant angular velocity. The detailed development of the set of equations describing the processes in this situation can be found in many books, e.g. in monographs by Chandrasekhar (1961) and Greenspan (1968). We describe here the main results necessary for further material presentation. The momentum for a fluid particle is given by:

$$\frac{\partial \overline{V}}{\partial t} + \left(\overline{V}\nabla\right)\overline{V} = -\nabla\left(\frac{P}{\rho} - \frac{1}{2}\left[\overline{\Omega}\times\overline{r}\right]^2\right) + \nabla\left(\nu\nabla\overline{V}\right) - 2\left[\overline{\Omega}\times\overline{V}\right] + \overline{F} \tag{1.1}$$

Here $\overline{V}$ is the fluid particle velocity vector, P the pressure, ρ the density of the particle, $\overline{r}$ the radius-vector of the particle from the rotation axis, $\overline{\Omega}$ the angular rotation vector, ν the kinematic viscosity of the fluid and $\overline{F}$ the vector of mass forces, e.g. gravity. The Coriolis force and centrifugal forces are presented here explicitly.

The momentum equation is supplemented by the continuity equation

$$\frac{\partial \rho}{\partial t} + \nabla\rho\overline{V} = 0 \tag{1.2}$$

and by the energy equation (see also Landau and Lifshiz, 1989)

$$\rho c_p\left(\frac{\partial T}{\partial t} + \overline{V}\nabla T\right) = \nabla\left(\kappa\nabla T\right) + \rho\left(\varepsilon + q\right) \tag{1.3}$$

where c_p is the specific heat capacity of the fluid, T its temperature, κ the heat conductivity. The rate of the kinetic energy dissipation into heat is determined by the Stokes formula

$$\varepsilon = \nu\frac{\partial V_i}{\partial x_k}\left(\frac{\partial V_i}{\partial x_k} + \frac{\partial V_k}{\partial x_i}\right) = \frac{1}{2}\nu\left(\frac{\partial V_i}{\partial x_k} + \frac{\partial V_k}{\partial x_i}\right)^2 = \tau_{ik}\frac{\partial V_i}{\partial x_k} \tag{1.4}$$

where i, k =1,2,3 and for repeating indices the summation is assumed, τ_{ik} is the viscous stress tensor, q is the rate of heating of unit mass from internal sources, e.g.

from chemical reactions or radiation absorption. If the fluid is inhomogeneous in its chemical composition, as e.g. salty sea water, then additional equations for the components should be considered. Virtually all convective motions in nature and most of those in technological applications have velocity scales much less than the speed of sound. Then (see Landau and Lifshitz, 1989) the density variations can be neglected everywhere and instead of Eq.(1.2) we simply have

$$\nabla\overline{V} = 0$$

This condition means that the density variations $\delta\rho/\rho_o$ are small; then we may assume a constancy of the kinetic coefficients ν and κ. All these conditions constitute the so called Boussinesq-type approximations for which the term $\rho\varepsilon$ in Eq.(1.3) is also small compared to other terms. In section 1.5 below we return to a consideration of the physical meaning of these assumptions in some more detail. In this case and in the absence of internal energy sources the equations (1.1) - (1.3) acquire the form

$$\frac{\partial\overline{V}}{\partial t} + \left(\overline{V}\nabla\right)\overline{V} = -\frac{\nabla P_1}{\rho_o} + \nu\nabla^2\overline{V} - 2\left[\overline{\Omega}\times\overline{V}\right] + \left(1 + \frac{\delta\rho}{\rho_o}\right)\overline{F} \tag{1.5}$$

$$\nabla\overline{V} = 0 \tag{1.6}$$

$$\frac{\partial T}{\partial t} + \overline{V}\nabla T = k\cdot\nabla^2 T \tag{1.7}$$

where $k = \frac{k}{\rho c_p}$ is the thermodiffusivity of the fluid. In Eq.(1.5) the therm $P_1 = P - \frac{1}{2}\left[\overline{\Omega}\times\overline{r}\right]^2$ is the pressure modified by centrifugal force. Such a formulation is convenient when the force is much less than the gravity acceleration. The opposite case is considered in Chapter 6.

We consider only the case when the mass force is $\overline{F} = -\nabla\Phi = \overline{g}$, where Φ is the gravity potential and therefore $\overline{g}$ is the gravity acceleration. We assume that in the vertical component of Eq. (1.5) the static distribution of the systematic vertical pressure gradient is equilibrated by the gravity acceleration and we will consider the rest of the pressure changes to this hydrostatic distribution. Then Eq. (1.5) will be of the form

$$\frac{\partial\overline{V}}{\partial t} + \left(\overline{V}\nabla\right)\overline{V} = -\frac{\nabla P_1}{\rho_o} + \frac{\delta\rho}{\rho_o}\overline{g} + \nu\nabla^2\overline{V} - 2\left[\overline{\Omega}\times\overline{V}\right]. \tag{1.8}$$

The equation system (1.6) - (1.8) forms a basis to study convective motions in a rotating fluid when the rotation is not too strong.

One of the most important features of the motions in the rotating system is established by so-called Proudman-Taylor theorem. To prove it we take the curl of Eq.(1.8) and assuming the vectors $\overline{g} = (0,0,g)$ and $\overline{\Omega} = (0,0,\Omega)$ are constant, we obtain for $\varpi = rot\overline{V}$ and $\nu = 0$

$$\frac{\partial\varpi}{\partial t} + rot\left[\overline{\varpi}\overline{V}\right] = \left[\nabla\frac{\delta\rho}{\rho_o}\times\overline{g}\right] + \left(\overline{\Omega}\nabla\right)\overline{V} \tag{1.9}$$

for an inviscid fluid. For stationary slow motions when non-linear effects are small, we have a zero left-hand side of Eq.(1.9). If we take the vertical (third) component of the rest, the term $\left[\nabla \frac{\delta \rho}{\rho} \times \overline{g}\right]$ is identically zero because $\overline{g} = (0, 0, g)$ and we have

$$\Omega \frac{\partial V_3}{\partial x_3} = 0 \tag{1.10}$$

If $\Omega \neq 0$ then $\frac{\partial V_3}{\partial x_3} = 0$, i.e. the vertical velocity component should not change in the direction of the rotation vector and if it is zero at some surface it should be absent throughout the fluid. The slow motions in an inviscid fluid should be two-dimensional. This was splendidly demonstrated by G.I. Taylor who slowly moved a obstacle at the bottom of a rotating vessel and observed a stagnant column above it. Such columns have received the name of Taylor column.

The Proudman-Taylor theorem does not hold in the presence of viscosity. This means that in a rotating fluid layer vertical motions can exist only due to the viscosity and should be observed only in regions with strong shears where the viscous forces gain their importance. Laboratory experiments do exhibit striking and beautiful examples of various flow patterns in convective rotating fluids (see sections 2.5-2.7 and others).

1.2 The basic non-dimensional parameters

Let us choose scales of dimensional variables in the form

$$\overline{V} = U\,\overline{V'}, t = \tau\, t', \nabla = L^{-1}\,\nabla', P = P_0\, P', T = \delta T\, T' \tag{1.11}$$

where U, τ, L, P_0, δT are scales of velocity, time, length, pressure and temperature. A concrete choice of these scales is determined by the boundary and/or initial conditions of the problem, or simply by our wish to estimate relative values of different terms of our system of equations. We also introduce unit vectors $\overline{k}$ and $\overline{n}$ in the directions of vectors $\overline{g}$ and $\overline{\Omega}$ (which can coincide). Introducing the scales (1.11) into Eqs. (1.8), (1.6) and (1.7), after straightforward manipulations we obtain the following non-dimensional equations (below we omit primes in all dimensionless variables):

$$St^{-1} \frac{\partial \overline{V}}{\partial t} + \left(\overline{V}\nabla\right)\overline{V} = -Eu^{-1}\nabla P + (B^*)^{-1}\,\overline{k} - Ro^{-1}\left[\overline{n} \times \overline{V}\right] + Re^{-1}\nabla^2\overline{V} \tag{1.12}$$

$$\nabla\overline{V} = 0 \tag{1.13}$$

$$St^{-1}\frac{\partial T}{\partial t} + (V\nabla)\,T = Pe^{-1}\nabla^2 T \tag{1.14}$$

We have here six non-dimensional numbers:

$$St = \frac{U\tau}{L}, \text{ the Strouhal number,} \tag{1.15}$$

$$Eu = \frac{\rho U^2}{P}, \quad \text{the Euler number} \tag{1.16}$$

$$B^* = \frac{U^2}{Lg\left(\frac{\delta\rho}{\rho}\right)}, \text{the buoyancy parameter,} \tag{1.17}$$

$$Ro = \frac{U}{2\Omega L}, \quad \text{the Rossby number,} \tag{1.18}$$

$$Re = \frac{UL}{\nu}, \quad \text{the Reynolds number,} \tag{1.19}$$

$$Pe = \frac{Ul}{k}, \quad \text{the Peclet number.} \tag{1.20}$$

It is useful to note that

$$Pe = Re\frac{\nu}{k} = RePr \tag{1.21}$$

where

$$Pr = \frac{\nu}{k}, \quad \text{the Prandtl number,} \tag{1.22}$$

characterizing the molecular properties of the fluid studied.

The Strouhal number determines the degree of non-stationarity of the fluid motion, the Euler number characterizes the pressure and in our case of the Boussinesq approximation when the pressure is the hydrostatic one, $Eu = O(1)$. The Peclet number is analogous to the Reynolds number and both determine how the advective transport of heat or momentum relates to the molecular transport.

The Rossby number Ro serves as a basic characteristic of motions in rotating systems and measures the ratio of inertial forces (or momentum advection) and Coriolis forces. When Ro is large compared to unity the influence of rotation is small. When $Ro << 1$ the rotation is decisive in determining the flow regime and its patterns. In the Russian literature, it is usually called the Kibel number after I.A. Kibel, who introduced it in 1941 in his seminal work on numerical weather forecast.

The buoyancy parameter B^*, first introduced in this generalized form by Batchelor (1953), is the ratio between inertial and buoyancy forces. In paper, Batchelor established the relationship of his number with the Richardson number

$$Ri = g\frac{\frac{\partial\rho}{\partial z}}{\rho\left(\frac{\partial V}{\partial z}\right)^2} \tag{1.23}$$

which appears in problems of stratified fluid motion in the presence of gravity. The Richardson number is a localized similarity criterion because it usually depends on the vertical coordinate z. But if we consider a layer with thickness L and assume within it the linear altitude dependence of density and velocity, then the Richardson number will coincide with the inverse buoyancy parameter B^*. For obvious reasons some authors call $(B^*)^{-1}$ the global Richardson number Ri_0 as is characterizes the whole layer. In general, hydrodynamics investigators usually use the Froude number

$$Fr = \frac{U}{\sqrt{Lg}} \tag{1.24}$$

and the internal Froude number

$$Fr_i = \frac{U}{\sqrt{Lg\left(\frac{\delta\rho}{\rho}\right)}} \tag{1.25}$$

Some preference to the buoyancy number B^* (1.17) follows from the following considerations. First, the uniformity of Eqs.(1.12) and (1.14) where all the non-dimensional numbers are at denominators of corresponding terms. The smallness of any of these numbers means the importance of the corresponding term, or the process, for the flow studied. We have already mentioned that at Rossby number $Ro << 1$ the role of rotation is determining the flow patterns. At $Re << 1$, it is essential to take into account the viscous forces, etc. Exactly in the same manner at $B^* << 1$ one must pay attention to buoyancy forces. Second, the form (1.17) of the buoyancy parameter allows one easily to obtain, under other assumptions, many other forms of non-dimensional parameters characterizing in various situations the driving buoyancy forces. Such examples will be given in the next section 1.4.

The transformation from Eq.(1.8) to Eq.(1.12) was performed by using the scale of U^2/L, i.e. the scale of the inertial term. Depending on the problem under consideration, one can use any other term of Eq.(1.8). The non-dimensional parameters appearing in this case will be combinations of the numbers presented by Eqs.(1.15)-(1.20). If we use as reference the viscous term or Coriolis term then we obtain another set of non-dimensional numbers (see, for example, Hopfinger,1992). The ones most often used are the Ekman number

$$Ek = \frac{Ro}{2\,Re} = \frac{\nu}{2\Omega L^2} \tag{1.26}$$

and the Prandtl number. Often, instead of the Ekman number its doubled inverse square is used, the Taylor number

$$Ta = \frac{4\Omega^2 L^4}{\nu^2} = Ek^{-2} \tag{1.27}$$

In dynamical meteorology, sometimes the Burgers number is used defined by

$$Bu = \frac{Ro^2}{B^*} = \frac{g\,\delta\rho}{4\rho\Omega^2 L} \tag{1.28}$$

In the Boussinesq approximation $\delta\rho/\rho = \alpha\,\delta T$, where α is the thermal expansion coefficient of the fluid.

1.3 Some other approximations

Some other simplifications of the equation system describing the motion of an inhomogeneously heated fluid can be carried out by using various approximations taking into account the specifics of the problem. For instance, when we study convective motions in a plane horizontal layer of the L with or without rotation, there is no systematic motion and initial velocity is zero when we start to consider the stability of a layer heated from below. But, following Lord Rayleigh, we can use the following scales: for the velocity $U = k/L$, time $\tau = L^2/\nu$, pressure $P_0 = \rho \nu k L^2$, temperature $L\nabla T$. Here ∇T is the initial temperature gradient and the buoyancy is caused only by the thermal expansion of the fluid with the equation of state $\rho = \rho_o\left[1 + \alpha\left(T - T_0\right)\right]$, where T_0 is some reference temperature. Then

$$St = Eu = Re = Pr^{-1}, \qquad Ro = \frac{Ek}{Pr}, \quad Pe = 1,$$

$$B^* = \frac{U^2}{\alpha g \nabla T L^2} = \frac{k^2}{\alpha g \nabla T L^4} = (Ra\ Pr)^{-1} \tag{1.29}$$

Here the buoyancy parameter B^* is replaced the well known Rayleigh number

$$Ra = \frac{\alpha g \nabla T L^4}{k\nu} = (B^*\ Pr)^{-1} \tag{1.30}$$

If the boundary conditions contain simply the temperature difference $\triangle T$ between lower and upper boundaries of the plane layer, then

$$Ra = \frac{\alpha g \triangle T L^3}{k\nu} \tag{1.31}$$

The linearized set of equations used to study the onset of convective instability in a plane horizontal rotating fluid layer now has the form

$$\frac{\partial \overline{V}}{\partial t} = -\nabla P + RaT\overline{k} - Ta^{\frac{1}{2}}\left[\overline{k} \times \overline{V}\right] + \triangle \overline{V}$$

$$\nabla \overline{V} = 0 \tag{1.32}$$

$$Pr\frac{\partial T}{\partial t} + \overline{V}\nabla T = \triangle T$$

Here $\triangle = \nabla^2$ and unit vectors $\overline{k}$ and $\overline{n}$ coincide. The determining nondimensional parameters are the Rayleigh, Taylor and Prandtl numbers. Some researchers use the Grasshof number instead of the Rayleigh number. Both numbers simply relate to each other:

$$Gr = \frac{Ra}{Pr} = \frac{\alpha g \triangle T L^3}{\nu^2} \tag{1.33}$$

Another type of approximation for rotating fluids is convenient for small Rossby numbers when basic temperature gradient is horizontal as in the case of laboratory experiments with rotating annuli which simulate the atmospheric general circulation. In case of horizontal temperature inhomogeneity, the motions exist in the fluid at arbitrary small deviations of the temperature at gravity equipotential surfaces. Most easily this can be seen from Eq.(1.9) in which, rotation or no rotation, nothing can compensate the term $[\nabla(\delta\rho/\rho)\times\overline{g}]$ if $\nabla(\delta\rho/\rho)\neq 0$. In this case still another velocity scale is chosen.

At $Ro << 1$ the so-called geostrophic approximation is used when the Coriolis force is equilibrated by buoyancy (remember that the buoyancy force in the Boussinesq approximation is the part of the pressure unbalanced by hydrostatics). Then $Ro = B^* << 1$ and

$$\frac{U}{2\Omega L}=\frac{U^2}{Lg\left(\frac{\delta\rho}{\rho}\right)},\ \text{or}\ U=\frac{g}{2\Omega}\left(\frac{\delta\rho}{\rho}\right)$$

Before this we had only one length scale L. For convection with the horizontal temperature difference, two length scales are important: the height of the fluid layer H and the horizontal distance L for which the horizontal temperature difference is prescribed. The fluid velocity can be also a function of the aspect ratio $\frac{H}{L}$. Usually, for the horizontal convection in the rotating fluid the velocity scale is chosen as

$$U_{tw}=\frac{g}{2\Omega}\left(\frac{\delta\rho}{\rho}\right)\frac{L}{H} \tag{1.34}$$

(the so-called thermal wind), the time scale is taken as

$$\tau=\frac{2\Omega}{g}\left(\frac{\rho}{\delta\rho}\right)\frac{H^2}{L} \tag{1.35}$$

and

$$St=1,\quad Ro=B^*=\frac{gL}{4\Omega^2H^2}\left(\frac{\delta\rho}{\rho}\right)=Ro_T \tag{1.36}$$

Ro_T is called the thermal Rossby number because the thermal wind enters its definition as a velocity scale. It plays about the same role in the horizontal convection as the usual Rayleigh number for the vertical convection. Both numbers are modified forms of the buoyancy parameter B^* and represent the motion driving force due to temperature or density inhomogeneity in a fluid under gravity force.

Eq.(1.12) for the horizontal convection now takes the form

$$\frac{\partial\overline{V}}{\partial t}+\left(\overline{V}\nabla\right)\overline{V}=Eu^{-1}\nabla P+Ro_T^{-1}\left(\overline{k}-\left[\overline{n}\times\overline{V}\right]\right)+Re^{-1}\Delta\overline{V} \tag{1.37}$$

1.4 Boundary layers

Studies of large classes of hydrodynamic problems have shown that, depending on the value of non-dimensional parameters, the appearance, evolution and change of the flow regimes take place according to almost the same scenario. At small values of the non-dimensional parameter (Reynolds, Rossby, Peclet or a similar number for a particular problem) this parameter completely or largely determines the flow regime. At an increase of the parameters value its role is weakening and at large values its influence in the main volume of the fluid becomes negligible. A self-similar regime may develop there. All the influence of these parameters is observed only in narrow boundary layers where sharp changes of the flow parameters take place. The classic monograph by Schlichting (1951) is probably the best reading for detailed acquittance with the boundary layer concepts and details of both theory and experiments for the nonrotating case.

The simplest example of a boundary layer is the viscous layer forming at the flow of a non-stratified non-rotating fluid with a single determining parameter, the Reynolds number. For a semi-infinite plane in the laminar flow the boundary layer is formed thickness δ_ν is determined by $\delta_\nu = l\,Re^{-\frac{1}{2}}$, where l is the distance from the plane edge. This dependence follows from the simple fact that in the non-dimensional Eq.(1.12) (in our case $Ro \to \infty$, $B^* \to \infty$, $\delta t \to \infty$, stationary flow and $Eu = O(1)$) the Reynolds number is in the denominator of the term with the Laplacian and to make it homogeneous we have the transformation $x_i' = Re^{\frac{1}{2}}\,x_i$, where x_i is the coordinate along the plate.

If the temperature of the plane (or plate) is higher than in the fluid then alongside with the viscous layer a thermal boundary layer is formed and the main temperature drop, as can be seen from Eq. (1.14), will take place within this last layer δ_T whose thickness is determined by $\delta_T = l\,Pe^{-\frac{1}{2}}$. The thickness of the viscous and thermal boundary layers relate to each other in this particular case as $\frac{\delta_\nu}{\delta_T} = \mathrm{Pr}$.

In case of the free convection and in absence of external flow there is an internal velocity scale for a layer with depth H. We can use it to determine the internal Peclet number. In this case the thickness of the thermal boundary layer is $\delta_T = H\,Pe^{-\frac{1}{2}}$.

In a rotating fluid near the horizontal boundaries the momentum due to viscous forces is balanced by the Coriolis force. From this balance, we determine the so-called laminar Ekman layer with depth $\delta_E = H\,Ek^{\frac{1}{2}} = \left(\frac{\nu}{\Omega}\right)^{\frac{1}{2}}$. Outside of this layer in the main body of fluid, the influence of viscosity is, as a rule, negligible.

In the rotating fluid special boundary layers form near vertical walls first introduced by Stewartson (1957). One of them is proportional to $H\,Ek^{\frac{1}{3}}$ and its physical origin is due to the necessity to fullfil the boundary conditions for normal and vertical velocity components at the vertical walls. In this layer, the fluid coming from the horizontal Ekman layer at the bottom or upper lid of a vessel is changing its flow direction and starts to flow vertically. To fulfil the continuity equation and smooth sharp changes and discontinuity in the flow azimuthal component second Stewartson boundary layer is formed with thickness proportional to $H\,Ek^{\frac{1}{4}}$. A detailed analysis of the boundary layers in rotating fluids can be found in the books by Greenspan (1968), Fein (1978); shorter reviews were presented by Veronis (1970) and Read (1992).

Veronis systematized the results for all three boundary layers which are pre-

	Ekman layer	$Ek^{\frac{1}{3}}$ **layer**	$Ek^{\frac{1}{4}}$ **layer**
Thickness of the layer, in h units	$Ek^{\frac{1}{2}}$	$Ek^{\frac{1}{3}}$	$Ek^{\frac{1}{4}}$
Spatial scale	$\frac{\partial}{\partial z} = Ek^{-\frac{1}{2}} \cdot \frac{\partial}{\partial \zeta}$	$\frac{\partial}{\partial x} = Ek^{-\frac{1}{3}} \cdot \frac{\partial}{\partial \xi}$	$\frac{\partial}{\partial y} = Ek^{-\frac{1}{4}} \cdot \frac{\partial}{\partial \vartheta}$
Scales of flow variables	$u, v, p = O(1)$, $w = O(Ek^{\frac{1}{2}})$	$p, u = O(Ek^{\frac{1}{2}})$, $v, w = O(Ek^{\frac{1}{6}})$	$v = O(1)$, $u = O(Ek^{\frac{1}{2}})$ $w, P = O(Ek^{\frac{1}{4}})$

Table 1.1 Scales of length and of main flow variables in the rotating fluid boundary layers after Veronis (1970)

sented in the Table 1. Here u, v, w are velocity components, radial, azimuthal and vertical for a cylindrical vessel ξ, ϑ and ζ are corresponding coordinates.

For convective motions in rotating fluids, the thermal boundary layers will be determined by the value of the buoyancy parameter B^* (or related parameters like the Rayleigh or thermal Rossby numbers) and by the Peclet number with some internal velocity scale. In a convective fluid with no rotation, thermal boundary layers are formed of the same scales as for rotating non-stratified fluids (with some external forcing, like fluid sources and sinks): at the horizontal surface there is the thermal boundary layer of order $H\,Ra^{-\frac{1}{4}}$ thick and at vertical walls there are Stewartson-type layers proportional to $Ra^{-\frac{1}{6}}$ and $Ra^{-\frac{1}{8}}$. In convective flows of rotating fluids, viscous boundary layers supplement the thermal ones. Similarly, for rotating non-stratified fluids these will be the Ekman and/or Stewartson layers. Their thickness at horizontal boundaries will be proportional to $Ek^{\frac{1}{2}}$, or $Ta^{-\frac{1}{4}}$ and at vertical walls to $Ek^{\frac{1}{3}}$ and $Ek^{\frac{1}{4}}$, or $Ta^{-\frac{1}{6}}$ and $Ta^{-\frac{1}{8}}$.

1.5 The efficiency of a convective system

How much of the supplied heat fluid can transform into a kinematic energy is a fundamental question in studies of convective motions. In more precise words, what is the fraction of the supplied heat power that goes into the generation of kinetic energy of convection? We consider this problem as did McKenzie et al (1974) and Golitsyn (1980b).

In the statistically steady state the average rate of the kinetic energy generation is equal to its rate of dissipation ε defined per unit mass. If M_0 is the total mass of the fluid in a vessel and Q is the total heat power supplied, then the efficiency we are discussing can be defined as

$$\eta = \frac{\varepsilon M_0}{Q} \tag{1.38}$$

Let us consider a volume **V**of a fluid confined by a surface **S**, where the fluid velocity or tangential stress τ_{ij} is equal to zero. Then the total energy balance in the volume by integrating Eq.(1.3) over the volume can be written using the Gauss theorem that $\int_{\mathbf{V}} \nabla^2 d\mathbf{V} = \int_{\mathbf{S}} \nabla \, d\mathbf{S}$ as

$$\int_{\mathbf{S}} k\frac{\partial T}{\partial x_i}\, d\mathbf{S} + \int_{\mathbf{V}} (\varepsilon + q)\, d\mathbf{V} = 0 \tag{1.39}$$

This means that the total heat flow over the surface **S** is equal to the total heating of the volume **V** by the dissipation of the kinetic energy and internal heat sources of the fluid under consideration.

Consider now the energy equation in its form for the specific entropy s (see e.g. Landau and Lifshitz, 1989):

$$\rho T\left(\frac{\partial s}{\partial t} + u_i\frac{\partial s}{\partial x_i}\right) = \frac{\partial}{\partial x_i}\left(k\frac{\partial T}{\partial x_i}\right) + \tau_{ij}\frac{\partial u_i}{\partial x_j} + q \tag{1.40}$$

For a chemically homogeneous fluid and using the thermodynamic identities and specific volume instead of the density, we have

$$ds = \left(\frac{\partial s}{\partial T}\right)_P dT + \left(\frac{\partial s}{\partial P}\right)_T dP = \frac{c_p}{T}\, dT - \left(\frac{\partial V}{\partial T}\right)_P dP = \frac{c_p}{T}\, dT - \alpha \mathbf{V}\, dP \tag{1.41}$$

Using Eq.(1.41) we transform Eq.(1.40) into

$$\rho c_p\frac{\partial T}{\partial t} - \alpha T\frac{\partial P}{\partial t} + \rho c_p u_i\frac{\partial T}{\partial x_i} - \alpha T u_i\frac{\partial P}{\partial x_i} - \frac{\partial}{\partial x_i}\left(k\frac{\partial T}{\partial x_i}\right) = \tau_{ij}\frac{\partial u_i}{\partial x_j} + q \tag{1.42}$$

In the statistically stationary case and taking into account the energy balance given by Eq.(1.39), we obtain

$$\int_{\mathbf{V}} \rho c_p u_i\frac{\partial T}{\partial x_i}\, d\mathbf{V} = \int_{\mathbf{V}} \alpha T u_i\frac{\partial P}{\partial x_i}\, d\mathbf{V} + \int_{\mathbf{V}} \tau_{ij}\frac{\partial u_i}{\partial x_j}\, d\mathbf{V} \tag{1.43}$$

The left-hand side integral, due to the continuity equation, can be transformed first into a divergency form, then to a surface integral by the Gauss theorem and then found is to be equal to zero due to conditions for velocity at the surface **S**. As a result, we have

$$\int_{\mathbf{V}} \left(\varepsilon + \alpha T u_i\frac{\partial P}{\partial x_i}\right) d\mathbf{V} = \mathbf{0} \tag{1.44}$$

Eq.(1.44) can be interpreted as the full compensation of the total energy dissipation by the work of the pressure forces. The last can also be called the rate of the kinetic energy generation. Note that there is no trace of rotation in the energy equation because the Coriolis force does not do any work: the mixed product $\overline{V}\left[\overline{\Omega}\times\overline{V}\right]\equiv 0$.

In case of ordinary convection heated from below, the pressure gradient is mostly the hydrostatic gradient. Therefore, Eq.(1.44) can be rewritten as

$$\mathbf{E}=\int_{\mathbf{V}}\varepsilon\, d\mathbf{V}=\int_{\mathbf{V}}(\alpha\rho gTw)\, d\mathbf{V},\quad w=u_i\frac{g_i}{g} \tag{1.45}$$

Let us introduce the scale height

$$\widetilde{H}=\frac{c_p}{\alpha g} \tag{1.46}$$

which characterizes the scale of the vertical density changes in the gravity field. For a perfect gas when $\alpha T\equiv 1$, it is the height of the adiabatically stratified fluid layer in which $\frac{ds}{dz}=0$ at the upper boundary of the layer (where also $T=\rho=0$.)

Assuming statistical homogeneity of motions in the layer in the horizontal direction for the unit area of the layer boundary, we can write

$$\mathbf{E}_1=\frac{1}{\widetilde{H}}\int\langle\rho c_p T\rangle\, dz,$$

where angular brackets mean statistical averaging. Here, and in (1.46), we also neglect the variations of the thermal expansion coefficient α with temperature and height, which complicates the algebra for any equation of state other than for perfect gas. If the Peclet number $Pe>>1$, the heat flux $f=\langle\rho c_p wT\rangle=const$ and is transported mainly by the motions, i.e. $\mathbf{E}_1=f\frac{H}{\widetilde{H}}$. Thus, the efficiency of convection in converting the supplied heat into the kinetic energy is

$$\eta=\frac{\mathbf{E}_1}{f}=\frac{H}{\widetilde{H}} \tag{1.47}$$

For the plane horizontal convecting layer, this energy transformation can be done by an equal consideration, taking into account explicitly the heat conduction (Golitsyn,1979,1980). The answer is

$$\eta=\frac{H}{\widetilde{H}}\left(1-Nu^{-1}\right) \tag{1.48}$$

where the Nusselt number characterizes the actual heat flux by convection to the one which would be caused by molecular heat conduction at the same temperature difference but with the fluid at rest. For the fluid layer with the depth H the Nusselt number is

$$Nu=\frac{fH}{\rho c_p k\,\Delta T} \tag{1.49}$$

When $Nu >> 1$, the expression (1.48) tends to (1.47). But even for rather well developed convection the efficiency in form (1.48) could be more appropriate than its asymptotic form (1.47) (see Golitsyn 1979).

1.6 Analogies between rotating and stratified fluids

Studying the flow of rotating fluid, one should pay attention to a comparison of the rotation and density stratification of the flow patterns in both cases. The physical basis for an analogy of motion types in these two cases lies in the existence in either case of a certain direction along which the motions are restricted. In rotating fluids, there is the direction of the angular velocity vector $\overline{\Omega}$. The Taylor-Proudman theorem forbids the motion along the rotation axis in an inviscid fluid with a boundary perpendicular to $\overline{\Omega}$. In a viscous fluid, such motions arise only in regions with strong shears, i.e. in some concentrated areas. In the stably stratified fluid there is the direction of the gravity acceleration vector $\overline{g}$ along which the density stratification creates a force driving a fluid particle back to its density level.

The analogy of this two effects has drawn attention of many researchers. A very detailed review of such analogies including the consideration of weakly stratified fluids with strong rotation (a large class of problems of convection in rotating fluids belongs to this) and of strongly stratified fluids with a slow rotation has been carried out by Veronis (1970). He used the equations (1.6) - (1.8) for a plane horizontal layer of depth h, rotating with the constant rate Ω. The density stratification was caused only by the temperature difference $\triangle T$ at the horizontal boundaries. For convenience, in tracing the analogies we introduced the dimensionless temperature separated from the vertical temperature gradient δT. Then, the energy equation (1.7) can be written as

$$\frac{2\,\delta T}{\tau}\frac{\partial T}{\partial t} + \frac{2\triangle T\,U}{h}\left(\overline{V}\nabla T\right) + \frac{4\,\delta T\,U}{h} w = \frac{2\triangle T\,k}{h^2}\cdot\nabla^2 T \qquad (1.50)$$

The velocity vector here has the horizontal and the vertical components $\overline{V}$ and w. For the case of the weak stratification and strong rotation, Veronis called this system $\mathbf{\Omega}_S$ when the geostrophic balance is assumed. The time scale is chosen as Ω^{-1} and pressure scale as $(\rho\Omega hU)$. From the momentum equation

$$\frac{2\alpha g\triangle T}{\Omega U} = \frac{2U\,\delta T}{\Omega h\triangle T}, \qquad or \qquad \frac{U}{\triangle T} = \sqrt{\frac{\alpha g h}{\delta T}} \qquad (1.51)$$

After the substitution into the equations (1.7) and (1.8) of the velocity scale, we obtain

$$\frac{\partial\overline{V}}{\partial t} + E^*\left(\overline{V}\nabla\right)\overline{V} + 2\left[\overline{k}\times\overline{V}\right] = -\nabla P + 2F^*T\overline{k} + Pr\,E_0\nabla^2\overline{V} \qquad (1.52)$$

$$\frac{\partial T}{\partial t} + E^*\left(\overline{V}\nabla T\right) + 2F^*w = Pr\,E_0\nabla^2 T \qquad (1.53)$$

where

$$F^* = \sqrt{\frac{\alpha g\,\delta T}{2}}, \quad E_0 = \frac{\nu}{2}\sqrt{Pr}, \quad E^* = \frac{U}{\Omega h} = 2Ro. \tag{1.54}$$

For strong stratification and slow rotation, the $\mathbf{S}_\Omega$ terms are important which conserve the hydrostatic approximation and oscillations caused by density inhomogeneity. This dictates the choice of the following scales for time, velocity and pressure:

$$\tau = \sqrt{\frac{h}{\alpha g\,\triangle T}}, \quad U = \sqrt{\frac{\alpha g h\,\delta T}{\triangle T}}, \quad P = \alpha g h \rho\,\delta T \tag{1.55}$$

and the equations appears as

$$\frac{\partial \overline{V}}{\partial t} + E^*\left(\overline{V}\nabla\right)\overline{V} + \frac{2}{F^*}\left[\overline{k}\times\overline{V}\right] = -\nabla P + 2T\overline{k} + PrR\nabla^2\overline{V} \tag{1.56}$$

$$\frac{\partial T}{\partial t} + E^*\left(\overline{V}\nabla\right)T + 2w = Pr^{-\frac{1}{2}}R\nabla^2 T \tag{1.57}$$

Parameter E^* for both systems is a measure of non-linearity. Let us show that for two-dimensional motions there exist an exact analogy between systems $\mathbf{\Omega}_S$ and $\mathbf{S}_\Omega$. For this we write the vector equation (1.52) on its components along $\overline{g}$ and $\overline{\Omega}$. Then we have

$$\frac{\partial u}{\partial t} + E^*\left(\overline{V}\nabla\right)u - 2v = -\frac{\partial P}{\partial x} + PrE_0\nabla^2 u \tag{1.58}$$

$$\frac{\partial v}{\partial t} + E^*\left(\overline{V}\nabla\right)v - 2u = -\frac{\partial P}{\partial y} + Pr^{\frac{1}{2}}E_0\nabla^2 v \tag{1.59}$$

$$\frac{\partial w}{\partial t} + E^*\left(\overline{V}\nabla\right)w - 2F^*T = -\frac{\partial P}{\partial z} + Pr^{\frac{1}{2}}E_0\nabla^2 w \tag{1.60}$$

$$\frac{\partial T}{\partial t} + E^*\left(\overline{V}\nabla\right)T + 2F^*T = Pr^{-\frac{1}{2}}E_0\nabla^2 T \tag{1.61}$$

$$\frac{\partial u}{\partial x} + \frac{\partial w}{\partial z} = 0 \tag{1.62}$$

For Pr $= 1$ the equation system in $\mathbf{S}_\Omega$ will be exactly the same as in $\mathbf{\Omega}_S$ with the following substitution of variables:

in $\mathbf{\Omega}_S$:	x	z	u	v	w	T	E_0	F^*
in $\mathbf{S}_\Omega$:	z	x	w	u	T	v	R	F^{*-1}

At Pr $\neq 1$, equations in $\mathbf{S}_\Omega$ have a difference with $\mathbf{\Omega}_S$ only in Eqs. (1.57) and (1.59): instead of $\mathrm{Pr}^{\frac{1}{2}}$ we have $\mathrm{Pr}^{-\frac{1}{2}}$.

In case of no stratification, the equation system (1.56) - (1.60) transforms into the system $\mathbf{S}_\Omega$ in which $F^* \to 0$ and the heat transfer equation is omitted. In the case of no rotation, the equation system (1.56) - (1.60) transforms into the system $\mathbf{\Omega}_S$ in which the angular momentum equation (1.57) is omitted and $F^* \to 0$.

Let us note the main physical analogies between systems $\mathbf{S}_\Omega$ and $\mathbf{\Omega}_S$.

(i) For stationary slow motions in the rotating fluid the Proudman-Taylor theorem is fulfilled as expressed by (1.10) which forbids the change of vertical velocity along the vertical axis. Similarly, in a stably stratified flow the variables do not change along the horizontal axis. In experiments with rotating fluids, we observe the Taylor columns over obstacles in which the fluid is stagnating relative to the surroundings and in stratified fluids we observe blocking, in which for motions along x of a cylinder with the axis along y we see formation of a horizontal layer pushed by the cylinder.

(ii) In both cases, when a fluid particle is shifted from an equilibrium position, a restoring force arises either in horizontal direction for rotation, or in the vertical for the case of stratification. Wave parameters for both types of waves (internal gravity and inertial waves are the same) the dispersion equation for waves of the form $\exp i\,(-\omega t + kx + nz)$ is

$$\omega = 2n\left(n^2 + k^2\right)^{-\frac{1}{2}} = \pm 2\cos\theta$$

where θ is the angle between z (or $\overline{\Omega}$) and the wave propagation direction for the $\mathbf{\Omega}_S$-system, and the angle between x and wave direction for the $\mathbf{S}_\Omega$-system.

(iii) Similarity in boundary layer structures both at horizontal and vertical walls in dependence of the basic similarity parameters - see section 1.4.

(iv) Stability properties of the flow for the two fundamental classes of motions: the Rayleigh-Benard convection and the Couette flow lead to identical stability equations and the same critical values of the basic Rayleigh and squared Taylor numbers. In the Rayleigh-Benard convection the plane layer with two horizontal rigid boundaries is heated from below and cooled from above. At not too large temperature differences $\triangle T$, a linear profile is formed and as it increases up to $Ra = \frac{\alpha g \triangle T\, h^3}{k\nu} = 1708$ the convective motions arise. The Couette flow is formed between two cylinders of different radii rotating with different angular rates along the same axis. At small $\triangle\Omega$, a linear change of fluid velocity is observed between the cylinders, but for sufficiently large $\triangle\Omega$, when $Ta = \frac{4(\triangle\Omega)^2 h^4}{\nu^2}$ becomes greater than 1708, the stability of this laminar flow pattern is lost and secondary flows in the form of circular rolls arise.

(v) There is a complete analogy between phenomena called spin-up and heat-up. Spin-up, according to Greenspan (1969) is the flow adjustment to small sudden changes of the rotation rate. Heat-up is a sudden change of the external temperature gradient by a small value.

Similar analogies can be traced out for systems $\mathbf{S}_\Omega$ and $\mathbf{\Omega}_S$. In this case, the analogies would be between rotating heat-up and stratified spin-up, between stratified Couette flow and rotating Rayleigh-Benard convection (see Veronis, 1970).

All these analogies are for two-dimensional and/or axisymmetric flows. In case of three-dimensional flow the analogies do not exist strictly speaking. However, in this cases of quasi-two-dimensional flows one should notice these analogies because they may be realized to a certain extent.

As a rule, the rotation, similarly to the stable stratification, increases the flow stability. However, situations exist when the rotation may cause changes of stability zones in a parameter space (Boubnov ,1980) or lead even to destabilization of flows (Vladimirov,1985) depending on the values of external parameters governing the flow patterns.

Chapter 2
PLANE HORIZONTAL HOMOGENEOUS LAYER

2.1 Convective motions in non-rotating fluids

Studies of convective motions in a plane horizontal layer with and without rotation have been considered in some detail in the monographs by Chandrasekhar (1961), Hopfinger (1992) and in the review by Yavorskaya and Belyaev (1985). In this section 2.1 we present some results of studies of convective motions in fluids with no rotation which are necessary for comparisons with results for rotating fluids. It is also necessary for answering the principal question: when is rotation important and when can it be neglected? Note that all experiments on the Earth are carried out in a rotating coordinate system and the value of the Taylor number for the water layer 20 cm deep for the Earth's rotation rate $\Omega = 7.3 \cdot 10^{-5}\, s^{-1}$ is of order unity.

Methods for convection studies in both cases of $\Omega = 0$ or $\Omega \neq 0$ are similar. For finding conditions for the onset of convective motions a linearized problem of the stability of a fluid at rest is considered. As is known (Chandrasekhar, 1961), the solution of this problem is degenerated and to determine the size and form of the appearing disturbance a study of non-linear effects is required. For the plane fluid layer heated from below it has been shown that the sole stable solution has the form of two-dimensional rolls. These rolls for a non-rotating fluid are subject to various types of instabilities depending on the value of the Prandtl number and on $Ra - Ra_{cr}$, i.e. how much the actual Rayleigh number exceeds the critical Rayleigh number of the convection onset.

For the three-dimensional case the convection is realized usually in the form of hexagons, first described by Benard (1900,1901).

The basic equation system for rotating fluids has been described by Eqs. (1.32). For $Ta = 0$, it transforms into a system first studied by Lord Rayleigh (1916) by the usual perturbation method, or by normal modes. In the initial state the fluid is at rest and in addition to this mechanical equilibrium, there is a thermal equilibrium when the temperature is linear with vertical coordinate z and does not change horizontally. Introduce small perturbations in dynamical variables and temperature from this state of equilibrium. Disregarding non-linear terms we obtain the system for perturbed values:

$$\frac{\partial \overline{V}}{\partial t} = -\nabla P + Ra\, \overline{k}\, T + \nabla^2 \overline{V} \tag{2.1}$$

$$\nabla \overline{V} = 0 \tag{2.2}$$

$$Pr \frac{\partial T}{\partial t} - \overline{V} \times \overline{k} = \nabla^2 T \tag{2.3}$$

This system is solved by expansion on normal modes and it determines the critical Rayleigh number (Ra_{cr}) when convection is arising. We shall carry out this procedure in detail for the rotating fluid layer. The values of Ra_{cr} together with the corresponding wave number a_{cr} are obtained for the temperatures prescribed at the lower and upper boundaries for various conditions on the velocity. The simplest case solved analytically to the end is when both boundaries are free. Then $Ra_{cr} = \frac{27\pi^4}{4} \approx 657.5$ and $a_{cr} = 2.22$. When the lower boundary is rigid and the velocity vector $\overline{v} = 0$ at $z = 0$ due to viscosity, then the values are obtained by solving a transcendental equation and $Ra_{cr} = 1100.7$ and $a_{cr} = 2.68$. For both rigid boundaries $Ra_{cr} = 1707.8$ and $a_{cr} = 3.12$. Note that the case with free boundaries is less stable than with rigid boundaries, i.e. the motions arise at smaller values of the Rayleigh number. Experiments on the onset of convection confirm well the linear stability analysis.

The patterns of arising convective flows can not be described in a purely linear approach. This case requires considering finite amplitude motions and taking into account non-linear effects. This can be done in two ways. First, one can consider stationary flow with prescribed form and scale and then the fine structure is determined together with some integral characteristics. Second, a non-stationary problem is solved at prescribed values of Ra and Pr and the preferred mode of convective motions is determined. In the first case the solution is usually expanded to degrees of a small parameter - the amplitude of motions near the critical value of the Rayleigh number. To determine the scale of motion a non-stationary problem with initial conditions is solved, or the stability of stationary supercritical flows is considered. The first approach is based on the method of interacting modes and is assuming the possibility of presentation of non-stationary motions in the region $Ra > Ra_{cr}$ as a superposition of some prime modes with time-dependent amplitudes. In the process of non-linear interaction, secondary modes are formed (and even higher order modes) and the mode interaction amplifies some modes and damps others. Efficiency in the use of the interacting modes method is largely determined by an intuitive choice of most effective modes and interactions among them.

The second approach consists in studying the stability of supercritical stationary flows. In this way it is shown that all spatial stationary structures (rectangular, hexagonal etc.) are found to be unstable. Only two-dimensional rolls are found to be stable at not too large supercritical Rayleigh numbers. With increasing Ra and in dependence on the wave length of perturbation and on the Prandtl number six types of instabilities for rolls have been found: sinusoidal, varicose, knotting, transversal, oscillatory and Ekhaus type.

Fig.1 presents the stability diagram for convective rolls in dependence on the Rayleigh and Prandtl numbers. This Figure reveals that the stable rolls exist only at not too large values of the supercritical Rayleigh numbers. Experimental studies by Willis et al (1972) confirm rather well these theoretical findings.

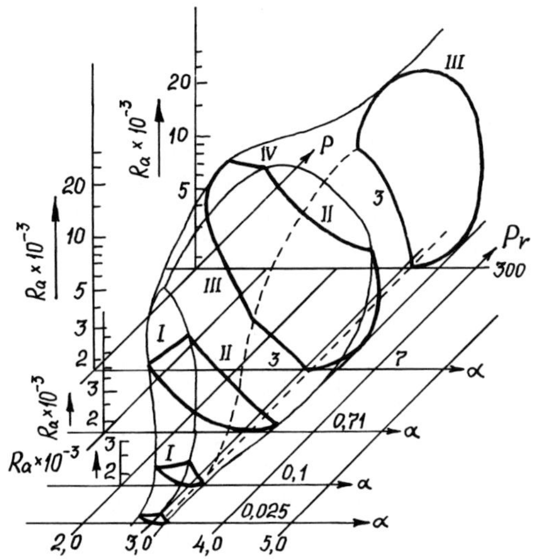

Figure 1: Stability regions of convective rolls in space (Ra,Pr, α). The stability boundaries corresponding to instability mechanism: I - oscillations; II - skewed varicose; III - Eckhaus, IV - knots. (After Busse, 1981, courtesy of Springer-Verlag).

2.2 The onset of convection in rotating fluids: Theory

The study of this problem is carried out by an analysis of the linearized equations (2.1)-(2.3) obtained from (1.32). The linearized momentum equation takes the form:

$$\frac{\partial \overline{V}}{\partial t} = -\nabla P + Ra\,\overline{k}\,\theta - Ta^{\frac{1}{2}}\left[\overline{n} \times \overline{V}\right] + \nabla^2 \overline{V}, \quad \overline{k} \equiv \overline{n} \tag{2.4}$$

Taking the curl of (2.4) and curl·curl of it, projecting the obtained equations on the vertical axis z, we obtain a system of two equations for the vertical velocity component w and the corresponding component $\xi = curl_z \overline{V}$:

$$\frac{\partial}{\partial t}\nabla^2 w = \nabla^4 w + Ra\,\triangle_1 T - Ta^{\frac{1}{2}}\frac{\partial \xi}{\partial z} \tag{2.5}$$

$$\frac{\partial \xi}{\partial t} = \nabla^2 \xi + Ta^{\frac{1}{2}}\frac{\partial w}{\partial z} \tag{2.6}$$

where $\triangle_1$ is the Laplacian on horizontal variables. The heat transfer equation is of the form

$$Pr\,\frac{\partial\theta}{\partial t} = w + \nabla^2\theta \tag{2.7}$$

We consider the motion in a plane horizontal layer and assume that small perturbations will expand on normal modes

$$\begin{aligned} w &= F(z)\exp\left[i\left(k_x x + k_y y\right) + \lambda t\right] \\ \theta &= \Theta(z)\exp\left[i\left(k_x x + k_y y\right) + \lambda t\right] \\ \xi &= Z(z)\exp\left[i\left(k_x x + k_y y\right) + \lambda t\right] \end{aligned} \tag{2.8}$$

Then the equation system (2.5) - (2.7) will have the following form

$$\left(D^2 - a^2\right)\left(D^2 - a^2 - \sigma\right)F - Ta^{\frac{1}{2}}\ Z = Ra\,\Theta\,a^2 \tag{2.9}$$

$$\left(D^2 - a^2 - \sigma\right)Z = -Ta^{\frac{1}{2}}\,F \tag{2.10}$$

$$\left(D^2 - a^2 - Pr\,\sigma\right) = -F \tag{2.11}$$

Here, the vertical coordinate z is normalized by the layer depth h

$$D = \frac{\partial}{\partial z},\ \frac{\partial}{\partial t} = -\lambda,\ \ \sigma = -\frac{h^2\lambda}{\nu},\ \ a = k_1 h,\ -k_1 = \frac{\partial^2}{\partial x^2} + \frac{\partial^2}{\partial y^2} = -\sqrt{k_x^2 + k_y^2}.$$

Multiplying Eq.(2.9) by $(D^2 - a^2 - Pr\cdot\sigma)(D^2 - a^2 - \sigma)$ and taking into account Eqs.(2.10) and (2.11), we can obtain one equation relative to the single variable F:

$$\begin{aligned} &(D^2 - a^2 - Pr\,\sigma)\left[(D^2 - a^2 - \sigma)(D^2 - a^2) + Ta\,D^2\right]F = \\ &-Ra\,a^2\,(D^2 - a^2 - \sigma)\,F \end{aligned} \tag{2.12}$$

The solutions depend, of course, on the types of the boundary conditions for velocity and temperature. Let us consider various boundary conditions in a sequence.

2.2.1 Free boundaries with fixed temperatures.

In this case the temperature perturbations and tangential velocity stresses are absent at both boundaries. Conditions for the velocity vertical component ξ and for temperature perturbations are

$$\xi = \frac{\partial^2\xi}{\partial z^2} = 0 \quad \text{at} \quad z = 0 \text{ and } z = 1$$

and for amplitude perturbations again at both boundaries:

$$F = \Theta = 0 \text{ and } DZ = 0, \qquad D^2F = 0 \tag{2.13}$$

Let us consider first the monotonic unstable mode or the stationary convection. In this case $\sigma = 0$. The solution of Eq.(2.12) is sought in the form $F = A \sin \pi n z$, where A is a constant and n is an integer. Then we have

$$Z = A\,Ta^{\frac{1}{2}} \frac{\pi n}{(\pi n)^2 + a^2} \cos \pi n z \tag{2.14}$$

The substitution of these solutions into Eq.(2.12) gives the characteristic equation

$$Ra = a^{-2} \left[\left(\pi^2 n^2 + a^2 \right)^3 + \pi^2 n^2\, Ta \right] \tag{2.15}$$

The minimal characteristic Rayleigh number is for $n = 1$ and, introducing a new variable $a^2 = \pi^2 x$ we now obtain the dependence of the Rayleigh number on the Taylor number and on the wave number $a = \pi\sqrt{x}$:

$$Ra = \pi^4 x^{-1} \left[(1+x)^3 + \frac{Ta}{\pi^4} \right] \tag{2.16}$$

The minimum of this expression is found ($\frac{\partial Ra}{\partial x} = 0$) to be for

$$2x^3 + 3x^2 = 1 + \frac{Ta}{\pi^4} \tag{2.17}$$

The solution of Eq.(2.15) does not depend on the Prandtl number for the monotonic mode as for the case of no rotation. But now the critical Rayleigh number is a function of both x and Ta. For large Taylor numbers the root of Eq.(2.16) tends to

$$x_{\min} = \left(\frac{Ta}{2\pi^4} \right)^{\frac{1}{3}} \tag{2.18}$$

and the corresponding values of Ra_{cr} and the minimal critical wave number tend to

$$Ra_{cr} = \frac{3\,(2\pi^4)^{\frac{1}{3}}\, Ta^{\frac{2}{3}}}{2} \approx 8.7\, Ta^{\frac{2}{3}} \tag{2.19}$$

$$a_{\min} = 2^{-\frac{1}{6}} \pi^{\frac{1}{3}} Ta^{\frac{1}{6}} \approx 1.3\, Ta^{\frac{1}{6}} \tag{2.20}$$

For $Ta = 10^6$ these expressions are valid with precision better than few per cent.

For other types of boundary conditions, it is not possible to find analytical expressions for critical values of Ra_{cr} and $a_{\min}$. However, Chandrasekhar (1953, 1961), using numerical variational methods, has calculated the critical curves for two rigid boundaries and for lower rigid and upper free surfaces for prescribed temperatures. These critical curves are presented at Fig.2 as **a,b,c** for ordinary convection and A,B,C - for oscillatory convection. Note some features of these curves. First, they are intersecting which means that the onset of the stationary convection depends on the type of the boundary conditions at large Taylor numbers, as opposite to the case of smaller Taylor numbers or no rotation at all. For non-rotating fluids, the rigid boundaries are more stable under temperature perturbations because they present stronger requirements to the constancy of the temperature at the layer boundaries. If the fluid is rotating, the Ekman layers arise at the rigid horizontal boundaries (see Chapter.1, section 1.4). At small rotation rates, these layers are thick and the convection starts without their influence. For large Taylor numbers the layers become very thin and determine to a large degree the onset of the convection in the fluid. Analyzing the results presented in Fig.2 one can see that the Ekman boundary layers for constant temperature on the boundaries are more unstable to the temperature perturbations than at the free boundary case. Note that the actions of Ekman layers and the free boundaries become equal, i.e. the corresponding curves **a** and **b** intersect at about the same values of the critical Rayleigh and Taylor numbers: $Ta = 3.30 \cdot 10^3$ and $Ra_{cr} = 3.25 \cdot 10^3$.

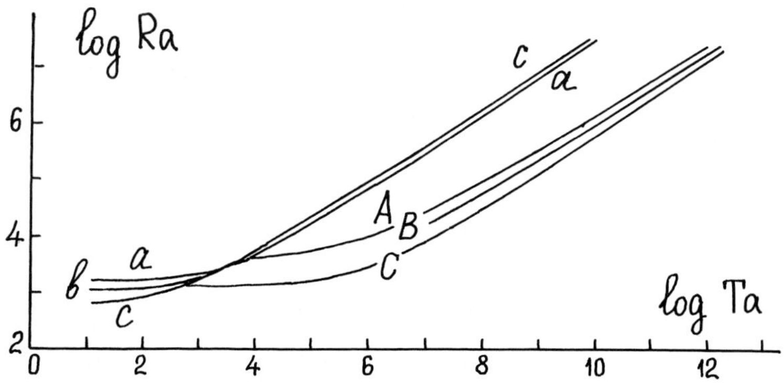

Figure 2: Stability diagram for the onset of convection for three cases: **a** and A - ordinary and overstability onset for two rigid boundaries, **b** and B - for one rigid other free boundaries, **c** and C - for two free boundaries. (After Chandrasekhar, 1961, by permission of Oxford University Press).

The second feature of Fig.2 is the presence of two well defined parts in the dependence $Ra_{cr}(Ta)$. For $Ta \leq 10^3$ the critical Rayleigh number changes only very little and for $Ta \geq 10^6$ the asymptotic behavior $Ra_{cr} \propto Ta^{\frac{2}{3}}$ is clear. In the first part, the rotation has almost no influence on the flow patterns and they are the same as in the case of no rotation, i.e. rolls, Benard cells, etc.

In rotating fluids, besides the monotonic mode there is also an oscillatory mode. It formally exists in the case $\Omega = 0$, but the critical Rayleigh number for it is higher than for the monotonic one, so it never develops in reality. For the solution of Eq.(2.12) this mode requires $\sigma \neq 0$. For both boundaries free we obtain the minimal

($n = 1$) value of the critical Rayleigh number which depends both on the Taylor and Prandtl numbers:

$$Ra_{cr} = 2\,(1 + Pr)\,x^{-1}\left[(1 + x)^3 + Pr^2\,(1 + Pr)^{-2}\,Ta\right] \tag{2.21}$$

where the critical value of the square of the non-dimensional wave number $x = \left(\frac{a}{\pi}\right)^2 = \left(\frac{k_1 h}{\pi}\right)^2$ is determined by a solution of the following equation

$$2x^3 + 3x^2 = 1 + Pr^2\left(1 + Pr^2\right)^{-1}\pi^{-4}\,Ta.$$

The corresponding frequency (squared) of neutral oscillations is

$$\omega_{cr}^2 = \frac{(1 - Pr)\,(Ta - Ta_1)}{(1 + x)\,(1 + Pr)}, \qquad Ta_1 = \frac{(1 + x)\,(1 + Pr)}{(1 - Pr)} \tag{2.22}$$

It is immediately clear that for $Pr \geq 1$ the oscillatory mode does not exist. More detailed but straightforward analysis for two free horizontal boundaries gives the critical value of the Prandtl number $Pr = 0.677$. At $Pr < Pr$, the oscillatory convection is more unstable than the stationary convection.

At large Taylor numbers, for the oscillatory convection the asymptotics for the critical Rayleigh number and corresponding wave number can also be found in case of two free boundaries:

$$Ra_{cr} = 3\,(1 + Pr)^{-\frac{1}{3}}\left(4\pi^2\,Pr^2\,Ta\right)^{\frac{2}{3}} \tag{2.23}$$

$$a_{\min} = [2\,(1 + Pr)]^{-\frac{1}{6}}\left(\pi^2\,Pr^2\,Ta\right)^{\frac{1}{6}}. \tag{2.24}$$

We see the same asymptotics at large Taylor numbers as for the stationary convection, only here the curves tend to their asymptotes at much larger values of Ta, while the corresponding Rayleigh numbers at the beginning of the asymptotic regimes are about the same (10^6) for both types of instability.

The critical curves for the onset of oscillatory convection for $Pr = 0.025$ (mercury) are also presented at Fig.2 for the three types of boundary conditions. However, here also, as for the stationary convection, the Ekman layers on the rigid surface change the ratio between critical Rayleigh numbers at a fixed Taylor number which causes the curves to get closer with increasing Taylor number.

2.2.2 *The constant heat flux at the layer boundaries*

In many technical and natural applications, at the boundaries of the plane layer the heat flux is prescribed, not the temperature. The solution of the problem for constant heat flux at the lower boundary $z = 0$ was considered by Boubnov and Senatorsky (1988). For constant temperature, the condition in terms of the function F from Eqs.(2.9) reads

$$\frac{\partial^4 F}{\partial z^4} - 2a^2\frac{\partial^2 F}{\partial z^2} = 0 \tag{2.25}$$

and for the constant heat flux it is

$$\frac{\partial^5 F}{\partial z^5} - 2a^2 \frac{\partial^3 F}{\partial z^3} = 0 \tag{2.26}$$

If both the temperature and heat flux are prescribed at the upper boundary, then we have for the free boundary at $z = 1$:

$$\frac{\partial^5 F}{\partial z^5} - 2a^2 \frac{\partial^3 F}{\partial z^3} + a^4 \frac{\partial F}{\partial z} = Bi \frac{\partial^4 F}{\partial z^4} \tag{2.27}$$

and at the rigid boundary

$$\frac{\partial^5 F}{\partial z^5} - 2a^2 \frac{\partial^3 F}{\partial z^3} = Bi \left(\frac{\partial^4 F}{\partial z^4} - 2a^2 \frac{\partial^2 F}{\partial z^2} \right) \tag{2.28}$$

where $-k\frac{\partial T}{\partial z} = T - T_\infty$, $Bi = h_0 h k^{-1}$ is the Biot number which determines the character of the heat transfer at a boundary: at $Bi = 0$ we have the fixed heat flux and for $Bi = \infty$ the temperature is fixed; here h_0 is the coefficient of heat transfer and T_∞ is the temperature of the surrounding medium. For these boundary conditions, the amplitude equations following Sparrow et al (1964) were solved by the method of Taylor series and critical parameters were determined for various boundary conditions. Let us numerate the first three variants of boundary conditions considered by Chandrasekhar (all were for fixed temperature): 1) both boundaries free, 2) both boundaries rigid, 3) the upper boundary free and lower rigid; the following variants are also possible for the rigid lower boundary: 4) fixed heat flux at both boundaries, 5) fixed heat flux at both boundaries, the upper one is also rigid, 6) the temperature is fixed at the lower boundary, the upper one is free with fixed heat flux, 7) same as in 6) but the upper boundary is rigid, 8) the fixed heat flux below and the free boundary above with fixed temperature, 9) same as 8) but with rigid upper boundary.

The critical curves for the onset of stationary convection and various types of boundary conditions, just numerated, and calculated by Boubnov & Senatorsky (1989a) are presented in Fig.3. Digits at the curves denote the variant numbers. Note some features of these curves. At the same dynamical boundary conditions, the curves shift practically parallel to each other (with precision better than 3 per cent) and the transition from one kind of the thermal condition to the other one is equivalent to multiplication of the critical Rayleigh number by a multiplier not depending on the value of the Taylor number (for $Ta > 10^6$). If we denote by indices T or f the conditions with fixed temperature or flux, then for the free upper boundary $Ra_{cr,T} = 3.38 \cdot Ra_{cr,f}$, and for the rigid upper boundary $Ra_{cr,T} = 2.49 \cdot Ra_{cr,f}$ (compare variants 3 and 4, 2 and 5). If the change of the thermal conditions takes place only at one lower boundary, then $Ra_{cr,T} = 2.09 \cdot Ra_{cr,f}$ for the free upper boundary (variants 4 and 6) and $Ra_{cr,T} = 1.80 \cdot Ra_{cr,f}$ for the rigid upper boundary (variants 5 and 7).

For the change of the thermal boundary conditions, the critical curves are getting closer and for variants with the temperature fixed at one of the boundaries one always observes the intersection of the stability curves. Fortuitously, or not, this intersection is taking place at approximately equal values of the Taylor and Rayleigh numbers, e.g. for variants 8 and 9 it is at $Ra_{cr} = 3282$ and $Ta = 3276$. For fixed heat flux, the curves do not intersect each other while changing the dynamical conditions. The pattern is the same as for the oscillatory instability, i.e. the difference between the critical values of the Rayleigh numbers is decreasing with the increase of the Taylor number, but no intersection of the curves occurs.

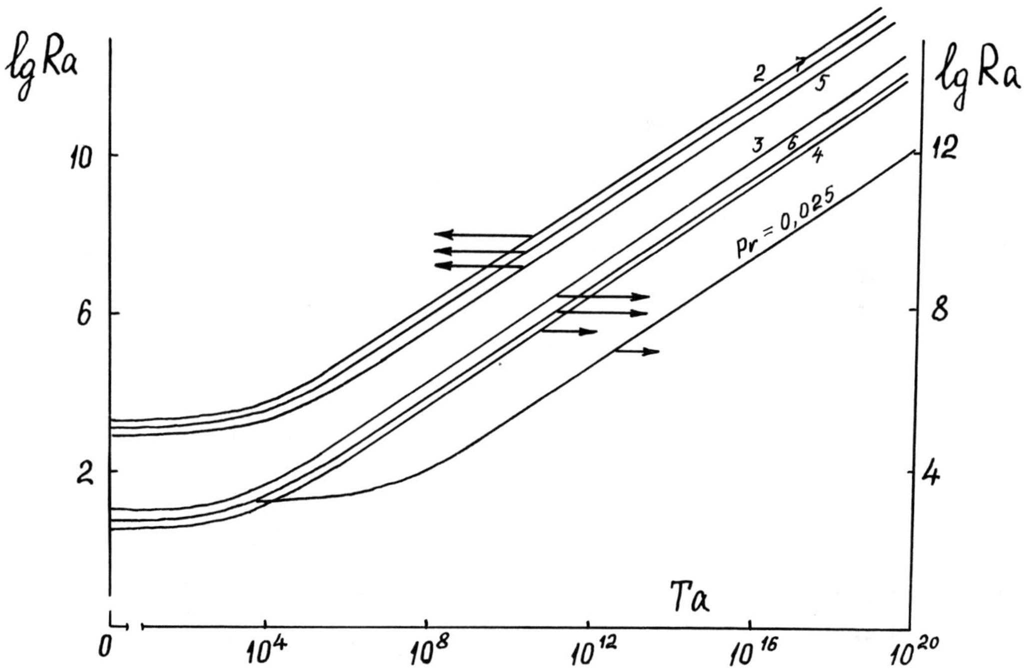

Figure 3: Dependence of critical Rayleigh number Ra_{cr} on Taylor number Ta for different boundary conditions (see text). Arrows indicate the different axis for the Rayleigh number. (After Boubnov & Senatorsky, 1989a).

At large Taylor numbers (practically at $Ta > 10^6$), all critical curves have the same asymptotes:

$$Ra_{cr} = k_R \cdot Ta^{\frac{2}{3}} \quad \text{and} \qquad a_{\min} = k_a \cdot Ta^{\frac{1}{6}} \tag{2.29}$$

The numerical values of the coefficients are presented in Table 2.1 for all 9 variants for two values of $Ta = 10^{10}$ and 10^{15} (the calculation have been also performed for $Ta = 10^{20}$ but the difference with the other cases was only in the fourth or fifth decimal place).

Variant	$k_R, 10^{10}$	$k_R, 10^{15}$	$k_a, 10^{10}$	$k_a, 10^{15}$
1	8.72	-	1.30	-
2	7.45	-	1.20	-
3	8.09	-	1.25	-
4	2.33	2.39	1.24	1.24
5	2.97	2.99	1.32	1.32
6	4.83	4.99	1.49	1.50
7	5.37	5.39	1.54	1.54
8	6.66	6.79	1.51	1.51
9	5.65	5.68	1.56	1.56

Table 2.1 Numerical coefficients in asymptotics (2.29) for $Ta = 10^{10}$ and 10^{15}(After Boubnov and Senatorsky, 1989a)

Beside the limiting cases of fixed temperature or heat flux, the problem was also considered for the dependence of Ra_{cr} on Ta for various values of the Biot number. This dependence for three different values of the Taylor number is presented at Fig.4. The similar dependence of the Rayleigh critical number on the Biot number is also observed for the no rotation case (see Sparrow et al,1964), only the distance between the curves is increasing with Ta.

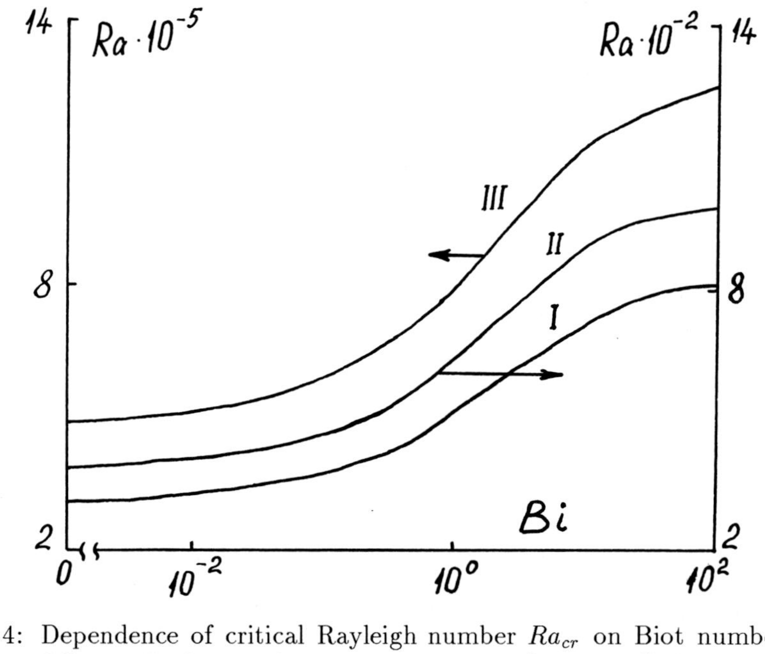

Figure 4: Dependence of critical Rayleigh number Ra_{cr} on Biot number Bi for three different Taylor numbers Ta : I - 0; II - 10^2; III - 10^8. (After Boubnov & Senatorsky, 1989a).

The oscillatory mode was also investigated by Boubnov & Senatorsky (1989a). The critical curves for its onset reveal similar changes. The values of the critical Prandtl numbers for the oscillatory mode prevail over the monotonic and were calculated for several variants of boundary conditions from the nine described above. If for variant 1 $Pr = 0.677$ as was calculated by Chandrasekhar, for variant 4 it is 0.343, for 5: 0.252, for 6: 0.365, for 7: 0.286. One of the critical curves for Pr= 0.025 (mercury) is also presented in Fig.3. At large Taylor numbers the critical curves for all types of boundary conditions for the oscillatory mode also behave as $Ta^{\frac{2}{3}}$.

2.2.3 Subcritical final amplitude convection

Up to now, we considered the linear problem which assumes the onset of convective motions with infinitesimal small amplitude. A study of convection with finite amplitude motions was first carried out by Veronis (1959, 1966) by means of the small parameter methods, i.e. the motions were assumed weakly non-linear. He considered a plane layer with free boundaries and the motions were two-dimensional. This assumption has an experimental justification since the convection starts in the form of two-dimensional rolls. Since $\frac{\partial}{\partial y} \equiv 0$, one can introduce the vorticity as $\eta = \frac{\partial u}{\partial z} - \frac{\partial w}{\partial x}$ and the stream function ψ, such that $u = \frac{\partial \psi}{\partial z}$, $w = -\frac{\partial \psi}{\partial x}$, wherefrom $\eta = \nabla^2 \psi$ and the system of momentum and heat transfer equations is transformed into:

$$\frac{\partial \eta}{\partial t} = J(\psi, \eta) + Pr\, Ta^{\frac{1}{2}} \frac{\partial u}{\partial z} - Pr\, Ra \frac{\partial T}{\partial x} + Pr\, \nabla^2 \eta \tag{2.30}$$

$$\frac{\partial w}{\partial t} = J(\psi, u) - Pr\, Ta^{\frac{1}{2}} \frac{\partial \psi}{\partial x} + Pr\, \nabla^2 w \tag{2.31}$$

$$\frac{\partial T}{\partial t} = J(\psi, T) - \frac{\partial \psi}{\partial x} + \nabla^2 T \tag{2.32}$$

Non-linear effects influence the temperature field through interaction between ψ and T , the horizontal velocity component through interaction between ψ and u. Due to this, in the temperature field a component $\sin 2\pi z$ type will be generated and in the horizontal velocity field a component will appear. The minimal system describing the finite amplitude convection can be represented as

$$\psi = A(t) \sin(\pi a x) \sin(\pi z) \tag{2.33}$$

$$T = B(t) \cos(\pi a x) \sin(\pi z) + C(t) \sin(2\pi z) \tag{2.34}$$

$$u = D(t) \sin(\pi a x) \cos(\pi z) + E(t) \sin(2\pi a x) \tag{2.35}$$

Substituting these expressions into the system (2.30) - (2.32) and equalizing the coefficients of the terms with $\sin(\pi a x)\sin(\pi z)$, $\cos(\pi a x)\sin(\pi z)$, etc., one can obtain a system of ordinary differential equations for coefficients A, B, C, D, E. A stationary solution of this system has been found, determining the values of the

Rayleigh numbers for which the finite solution exists at $Ra < Ra_{cr}$. For this solution the following relationships exist:

$$Ra = \left\{\left[\left(a^{-2} - Pr^2\right)\left(1 + a^2\right)^3 \pi^4\right]^{\frac{1}{2}} + Ta^{\frac{1}{2}}\right\} \tag{2.36}$$

$$Ta \geq Pr^4 \left(1 + a^2\right)^3 \pi^4 \left(a^{-2} - Pr^2\right) \tag{2.37}$$

which have a sense when $Pr < a^{-2}$. The minimal value of the Rayleigh number $Ra_f^{\min}$ for which the finite amplitude convection can exist is determined by the condition $a^2 = \left[1 \pm (1 - 3Pr^2)^{\frac{1}{2}}\right](3Pr^2)^{-2}$ which for small Prandtl number (for the minus sign) tends to $a^2 \to \frac{1}{2}$ and

$$a^2 \to \frac{1}{2} \quad \text{and} \quad Ra_{f.cr}^{\min} \to \left(\frac{3\sqrt{3}}{2}\pi^2 + Pr\, Ta^{\frac{1}{2}}\right)^2 \tag{2.38}$$

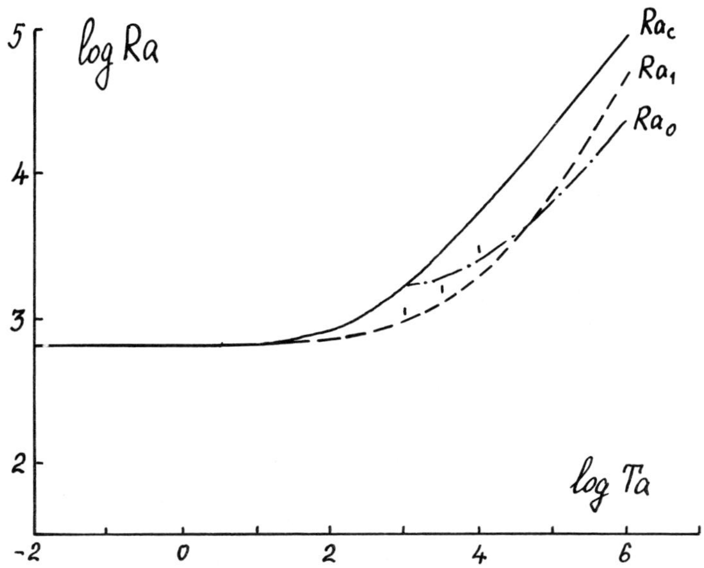

Figure 5: The minimum Rayleigh numbers Ra at which motion can be marginally maintained are plotted as function of Taylor number Ta with $Pr = 0.2$. The solid and dash-dot curves show the values at which linear theory predicts exchange of stability (Ra_c) and overstability (R_o). The dashed curve (Ra_1) shows the lowest values at which finite-amplitude motions can be maintained according to results from a minimal representation of the velocity and temperature fields. At $Ta = 10^3$, $10^{\frac{7}{2}}$ and 10^4 the minimum Ra for finite-amplitude instability will occur in the range designated by the small vertical lines. (After Veronis, 1968, courtesy of Cambridge University Press).

Ta	$\delta = 0.2$	$\delta = 0.5$	$\delta = 1$	$\delta = 2$	$\delta = \infty$
0	151.2	8.01	2.54	1.88	1.71
10^4	151.3	8.77	5.09	4.90	4.71
10^6	163.5	55.3	55.5	57.0	71.1
$1.96 \cdot 10^6$	174.9	83.7	83.4	85.7	111.7

Table 2.2 Critical values of the Rayleigh numbers $Ra_{cr} \cdot 10^{-3}$ for various Ta and aspect ratio $\delta = \frac{r}{h}$ for heat conducting side walls (After Buell and Catton,1983).

For very large Taylor numbers this expression asymptotically tends to $Ra_{f.c}^{\min} \rightarrow Pr^2\,Ta$, while for the linear stability $Ra_{cr} \rightarrow Ta^{\frac{2}{3}}$. Fig.4 presents critical stability curves obtained by Veronis for various types of convective motions; curve Ra_c is by Chandrasekhar for stationary convection for both boundaries free which does not depend on the Prandtl number, curve Ra_o is for the oscillatory convection (also by Chandrasekhar) at $Pr= 0.2$, curve Ra_1 is the critical curve by Veronis for finite amplitude convection also at $Pr= 0.2$. The vertical bars correspond to the critical values for the finite amplitude convection obtained by using more expansion terms and for the final results does not depend on the number of expansion terms chosen (after some number of the ones included). One can see a tendency to an asymptote at which the critical Rayleigh numbers come already parallel to the case of linear analysis dependence, though somewhat lower, which evidently means that the finiteness of the amplitude decreases the stability of the system. Unfortunately, the experiments are still too few and uncertain to confirm definitely these results.

2.2.4 The influence of the side boundaries

The theoretical results just described for the plane infinite horizontal layer served as a basis for comparison with experiments. To explain differences between theoretical and experiments results, a number of authors (Buell & Catton,1983, Herrmann & Busse,1993, Goldstein et al,1993) took into account the influence of the side walls on the onset of the convection. The motions were considered either in a infinite channel or in a rotating cylinder with finite fluid depth. Considering the motions in a cylinder and only axisymmetric disturbances, it was found that for them the critical Rayleigh number is always larger than for an infinite layer. The consideration of only the axisymmetric modes was chosen because for a non-rotating cylinder such modes are most often observed experimentally. Addition of no-axisymmetric modes by Buell & Catton (1983) has revealed that in dependence on the Taylor number and geometry parameters, there is no unique way of changes of at transition from infinite layers to finite radius cylinders. Table 2.2 and Table 2.3 presents some values of the critical numbers $Ra_{cr} \cdot 10^{-3}$ for various values of Taylor numbers and the aspect ratio $\delta = \frac{r}{h}$ where r is the radius of the cylinder with rigid isothermal upper boundaries.

Inspection of Table 2.2 and Table 2.3 shows that for small aspect ratios δ, the rotation influences the critical Rayleigh number only weakly and results of the linear analysis for the infinite layer can differ substantially from experiments at $\delta \leq 1$.

Ta	$\delta = 0.2$	$\delta = 0.5$	$\delta = 1$	$\delta = 2$	$\delta = \infty$
0	51.8	3.77	2.26	1.83	1.71
10^4	51.8	4.08	3.43	3.63	4.71
10^6	55.5	15.0	14.2	16.5	71.1
$1.96 \cdot 10^6$	58.9	19.8	18.7	22.6	111.7

Table 2.3 Critical values of the Rayleigh numbers $Ra_{cr} \cdot 10^{-3}$ for various Ta and aspect ratio δ for thermally insulated walls (After Buell and Catton,1983).

Goldstein et al (1993) continued consideration of non-axisymmetric azimuthal waves in a rotating cylinder. They found that instability is a Hopf bifurcation regardless of the Prandtl number and leads to precessing spiral patterns. In the frame of linear analysis, they consider two cases of boundary conditions A and B. A - free-slip, impenetrable, infinitely conducting horizontal plates at top and bottom, and rigid, impenetrable, insulating sidewall. This case could not be compared with the experiment, but it is mathematically convenient and allows a separation of variables in vertical and azimuthal directions. B - rigid boundaries everywhere and an insulating sidewall (temperature boundary conditions on the sidewall are not critical to methods used). It was founded that convective instability gives rise to two types of precessing patterns called rotating waves: fast and slow modes. Fast modes peak near the walls and are preferable for small to moderate aspect ratios. For large aspect ratios, the fast modes go unstable become slow modes provided the rotation rate is sufficiently slow. This mode is low near the sidewall and peaks in the inner part of the cylinder. The precession velocity of the slow modes vanishes as $\delta \to 0$ and they provide a connection between the results for a finite-aspect ratio system and the infinite layer.

Herrmann & Busse (1993) used asymptotic theory for a consideration, in the limit of large rotation rates, of the influence of the vertical sidewalls. The case of the stress-free upper and lower boundaries and the plane sidewalls was considered. It was supposed that boundary-layer phenomena were in terms of the distance from the sidewalls, and a finite curvature of the wall was not an essential ingredient of the problem. It was found that when the sidewall is insulating the critical Rayleigh number is proportional to the Taylor number at the power $\frac{1}{2}$ ($Ra_{cr} \propto Ta^{\frac{1}{2}}$) and in the case of infinitely conducting sidewalls the critical Rayleigh number grows as in the case of an infinite layer, but with a coefficient of proportionality $0.9086\pi^{\frac{4}{3}}$.

2.3 The onset of convection in rotating fluids. Experiments.

2.3.1 Stationary convection

Experiments on convection in rotating fluids has started along with theoretical studies. In the first half of the 1950s, Nakagawa and Frenzen (1955) described, alongside with their calculations of the stability curve and critical wavelength, the first experiments on testing them. The experiments were carried out with water ($Pr \approx 7$) with a free upper surface and a satisfactory agreement was obtained with theoretical results.

For determining the critical Rayleigh number, the so-called gradient method was used which is sometimes called Schmidt-Milverton principle. In this method the vertical temperature gradient within the layer is measured in dependence on the temperature difference at the layer boundaries. Increasing this difference, we proportionally increase the gradient up to the moment the motion will start. The further increase of the temperature difference $\triangle T$ leads to a slower increase of the temperature gradient in the central part of the layer and to a faster increase near the boundaries, due to the fact that near the boundaries the thermal boundary layers start forming. The value of the gradient change in dependence of $\frac{dT}{dz}$ on $\triangle T$ is taken as the critical temperature gradient which is used to calculate the critical Rayleigh number.

The measured dependence of Ra_{cr} for the case of the free upper surface is presented in Fig.6 where results by Nakagawa & Frenzen (1955) and Boubnov & Senatorsky (1989b) are also shown. A good agreement of the theory and experimental results is observed in a wide range of the Taylor numbers: from 10^6 to 10^{10} . One may note that at larger Taylor numbers experimental results start to deviate upward from the theoretical stability curve which might be, possibly, explained by the influence of the centrifugal effects at large rotation rates of the system.

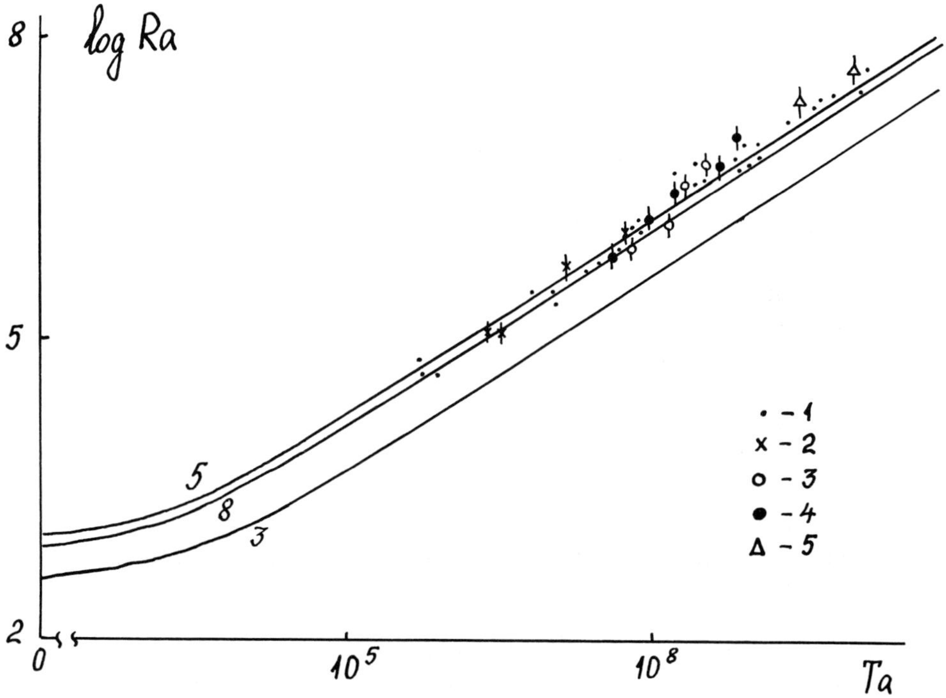

Figure 6: Critical curves for the convection onset for a free upper boundary: digits on the curves denote the variant numbers. Points: 1-Nakagawa & Frenzen (1955); 2-5 $h = 3, 5, 7$ and $10\,cm$. (After Boubnov & Senatorsky, 1989b).

However, the experiments with the upper free surface of the layer in a container do not in their geometry fullfil the conditions of the plane horizontal layer because at the large Ω the upper surface acquires a parabolic form. Therefore, more suitable for comparison with the theory are experiments with two rigid boundaries. The most detailed experiments for determining the critical Rayleigh number for the onset of convection were carried out by T.Rossby (1969) with water and mercury and a series of works by Lucas et al (1983) and Phontenhauer et al (1984,1987) who used the liquid ^4He as working fluid. For the determinations of the critical Rayleigh numbers in both these studies the "flux" method was used. In this method, the heat coming through the fluid is measured while increasing the values of Ra.

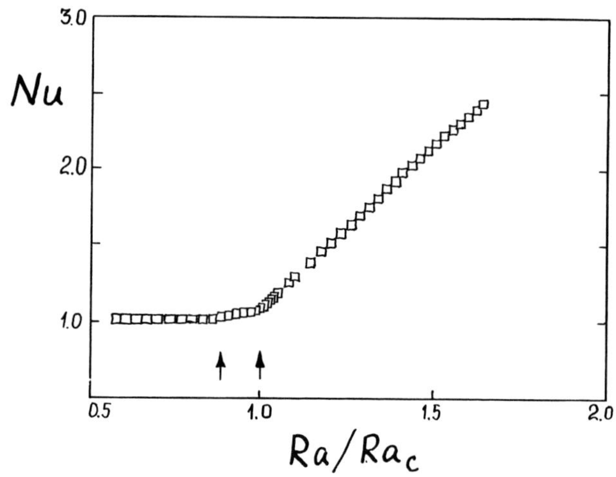

Figure 7: Dependence of Nusselt number on the relative Rayleigh number $(\frac{Ra}{Ra_c})$. The first break in the data occurs at Rayleigh number less than expected for linear convection. Ra_c, or the second break, occurs within expected uncertainty of the value predicted by linear theory. (After Lucas et al, 1983, courtesy of Cambridge University Press).

An example of such a dependence is shown in Fig.7. When $Ra < Ra_{cr}$ the motions are absent in the fluid and the heat flux is formed by purely molecular mechanisms when the Nusselt number $Nu = 1$. For $Ra \geq Ra_{cr}$, the Nusselt number becomes $Nu \geq 1$ caused by the appearance of the motions in the fluid. However, one can see in Fig.7 a small increase of the Nusselt number below the theoretical critical Rayleigh number which is shown by the left arrow. Results on measuring the values of the Ra_{cr} in dependence of Ta are given in Fig.8. At values of the Taylor number $Ta > 10^5$, the stationary convective motions arise at smaller Rayleigh numbers than those from the linear theory. For instance, at the experiments by T. Rossby for $Ta = 10^8$, the experimental value of Ra_{cr} is only about a third of the theoretical one. Similar divergence of the experimental results with theoretical ones at larger values of Ta was noted in the work with the liquid helium. It was found as in the experiments by T. Rossby that the difference of the experimental critical values starts from some critical value of the Taylor number Ta_{cr} which in its turn is determined by the vessel aspect ratio, the parameter $2\delta = \frac{D}{h}$, the ratio of the vessel diameter D to the depth of the fluid h. The critical values of the Taylor number for some values of δ determined by Phontenhauer et al (1987) are presented in Fig.8. One can also see from this figure that for the given value of the Taylor number, the difference between the experimental Ra_{cr} and the theoretical one for the infinite layer is decreasing with the growth of the aspect ratio.

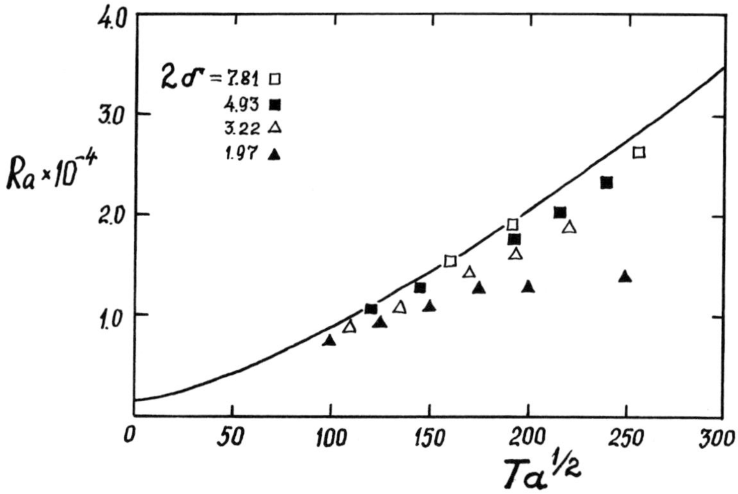

Figure 8 : Variation in critical Rayleigh numbers as function of dimensionless $Ta^{\frac{1}{2}}$ for different 2δ (after Phontenhauer et al, 1987, courtesy of Cambridge University Press).

To explain the existence of the convective motions at $Ra < Ra_{cr}$ an analysis was carried out by Phontenhauer et al (1987) of the main possible mechanisms of the instabilities. By testing the reproducibility of the experimental results at the values of the Prandtl number larger and smaller than the critical one, it was shown that the oscillatory mode does not play any significant role. The presence of the convective motions at $Ra < Ra_{cr}$ for different values of the Prandtl number (7 for water and 0.49-0.79 for ^4He at different temperatures) leads to the conclusion that finite amplitude disturbances of the two-dimensional flow are also insignificant because the deviation from the stability curve is observed for all values of the Prandtl number used while according to Veronis (1968) the results should differ principally at the transition for the value of the $Pr = \sqrt{2}$. Also, it was shown that centrifugal effects are of no importance at the onset of the convection. The last was proved by Phontenhauer et al (1987) who used the liquid helium which has the λ-point on its phase diagram at which the coefficient of the thermal expansion changes its sign. The same effects have been observed at either side of the λ-point. As the most plausible mechanism for explanation of the difference between linear theory for infinite layers and experiments, a mechanism of asymmetrical modes was proposed described by Buell & Catton (1983) which relates to the presence of the side wall, though the influence of the finite amplitude convection can not be fully excluded.

To consider the influence of the side wall, Zhong et al (1991) and Ecke at al (1992) made experiments for the water in a cylinder with $\delta = 1$. Thermally insulating plexiglas sidewall, sapphire top and copper bottom (both excellent thermal conductors) were used. They found waves in the rotating frame which take the form of m-armed spirals. Linear theory for case B (see section 2.2 and Goldstein et al, 1992) gives a good agreement with the experiment both in critical Rayleigh number and in the azimuthal mode number m for $Ta < 10^7$.

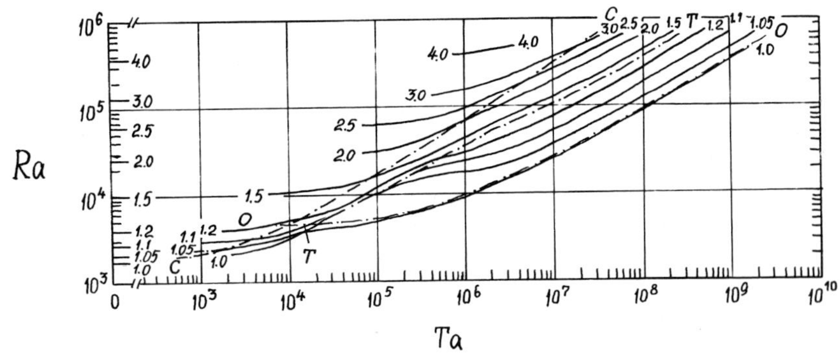

Figure 9: Stability curves and lines of constant Nusselt number as a function of the Rayleigh and Taylor numbers in mercury with Prandtl number $Pr = 0.025$. The dot-dashed lines O-O and C-C are the marginal stability lines according to Chandrasekhar (1961) for oscillatory and steady convection. The dot-dashed line T-T delineates the experimentally determined transition between oscillatory and irregular convection. (After Rossby, 1969, courtesy of Cambridge University Press).

2.3.2 *Oscillatory instability*

Experiments with this type of instability have been more definite. After theoretical discovery of this instability by Chandrasekhar and Nakagawa & Frenzen a number of investigators studied in much detail the convection in a rotating mercury layer ($Pr = 0.025$). They confirmed the main conclusions of the linear theory. The first experiments, e.g. the case of stationary convection in a rotating fluid, were carried out by Fultz et al (1954,1955). A special attention was paid to the mercury free upper surface. Because it is oxidized rather quickly, it behaves largely as a rigid upper surface with constant temperature. For determining the critical Rayleigh number, the gradient method was used.

Further results were obtained by T. Rossby (1969) which are represented in Fig. 9. In his detailed study, he found an excellent agreement of the theoretical and experimental results for $Ta > 10^5$. However, at smaller values of the Taylor number ($Ta < 2 \cdot 10^4$) the difference between theory and experiments is significant, which the author tries to explain by the finite amplitude perturbations. One of the interesting results of this work is the determination of the regions on the plane $Ra - Ta$ where the oscillatory instability exists in the form of regular or irregular oscillations. The boundary between these two regimes is presented by the line T-T in Fig.9. In similar experiments by Goroff (1960) published before T. Rossby, only regular oscillatory patterns were observed. The spectra for the oscillatory convection obtained by T.Rossby for the Rayleigh numbers significantly higher than the critical one but still smaller than the one for the monotonic instability, have no very sharp maximum which is actually quite broad in its dependence on the time frequency.

2.4 The structure of the convective motions and at small supercritical Rayleigh numbers

By small Rayleigh numbers we mean here $Ra < 10^6$. In the case of no rotation at $Ra > 10^6$, the convection is always turbulent. In the case of $\Omega \neq 0$ for $Ra < 10^6$ the motions exist for $Ta \leq 10^8$ (see the stability diagrams at Fig.2). At small values of Ra and not too large values of the Taylor numbers, the spatial structure of convective motions resembles much the one for the case $\Omega = 0$. A detailed consideration of convective motions of various forms was performed by Veronis and described by Chandrasekhar (1961). It was shown that stationary solutions of equations with finite but small amplitude are two-dimensional rolls and cells of square, rectanqular and hexagonal forms. The flow pattern for a hexagonal form is presented in Fig.10. The authors should note that similar hexagonal structures were occasionally observed in their experiments described below in sections 2.6-2.9, but existed only for a very short time. Further theoretical studies by Kupper & Lortz (1969,1970) have shown that the only stable solution which exists also at sufficiently small Taylor numbers is the convection along two-dimensional rolls. This solution is of the form:

$$u = -\frac{\pi}{a} W_0 \sin ax \cos \pi z$$

$$v = \frac{\pi Ta^{\frac{1}{2}}}{a(\pi^2 + a^2)} W_0 \sin ax \cos \pi z \tag{2.39}$$

$$w = W_0 \cos ax \sin \pi z$$

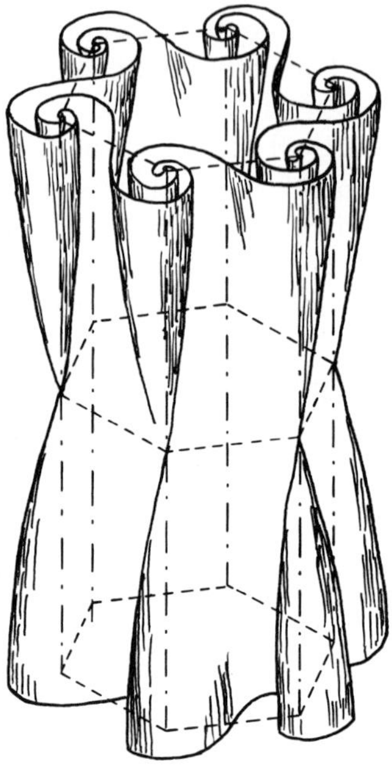

Figure 10: A perspective sketch of a hexagonal convection cell in rotating fluid. (After Veronis, 1959, courtesy of Cambridge University Press).

The difference with the case $\Omega = 0$ for the rolls is that in the latter case a fluid particle moves in the direction perpendicular to the axis of the rolls (see Fig.11) while in the rotating case the fluid particles move along a plane with the angle to the axis β for which

$$\tan\beta = \frac{u}{v} = -\frac{Ta^{\frac{1}{2}}}{\pi^2 + a^2} \tag{2.40}$$

Kupper & Lortz (1969) also showed that the wave number a_s in the plane of the fluid particle motion inclined at the angle β to the roll axis does not depend on the rotation rate and is equal to $a_s = \frac{\pi}{\sqrt{2}}$. These authors also studied the stability of the solution (2.39) for $Pr \rightarrow \infty$ and for the layer with free boundaries. They found that the rolls are stable only at weakly supercritical Rayleigh numbers and for $Ta < Ta_{cr} = 2285$ for these boundary conditions, and the critical angle χ_{cr} between the growing disturbances and the roll axis counted from the direction of the rotation rate $\overline{\Omega}$ is equal to $58° \pm 0.5°$. For $Ta > Ta_{cr}$, there is no stationary motions. The further development of these studies by Kuppers (1970) for the rigid boundaries and arbitrary Prandtl numbers showed that the critical Taylor number depends on the value of the Prandtl number. This dependence is presented in Fig.12, where the stable motions of finite amplitude subcritical convection are shown.

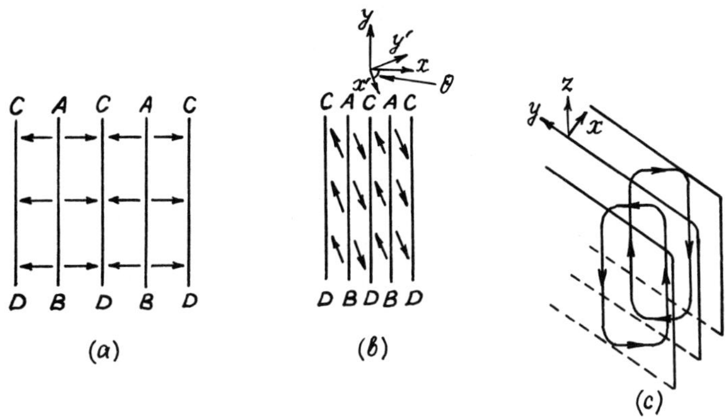

Figure 11: (a) A top view of two-dimensional rolls in a non-rotating fluid. The arrows indicate the direction of fluid motions (b) The same in a system rotating counterclockwise. (c) A perspective view of fluid particle motions in a roll when the fluid is rotated. (After Veronis, 1959, courtesy of Cambridge University Press).

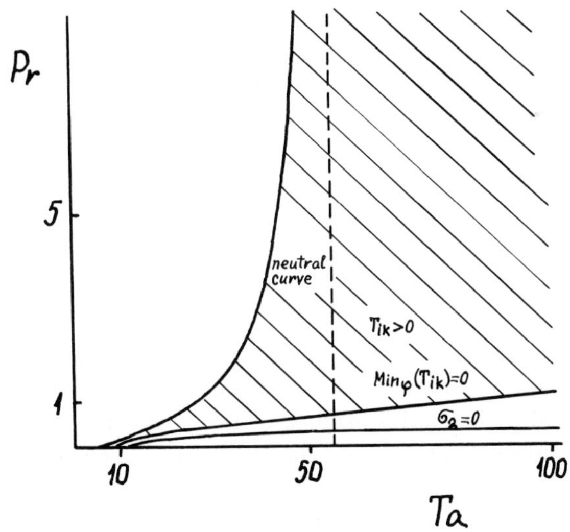

Figure 12: The stability diagram in (Pr, Ta) space. On the left side of the neutral curve, the rolls are stable; in the dashed region, no steady-state convection flow exists.(After Kupperts, 1970, courtesy of Phys.Lett.).

Busse & Clever (1979) were interested in the kind of motions that may exist in the parameter space where neither oscillatory convection nor the rolls can exist. They found that the only possibility is in the construction of aperiodic flow with non-linear dependence of its amplitude on time. Their solution included three two-dimensional perturbations with the same wave length and with wave vectors directed over an angle of 60° to the direction of the initial rolls.

A schematic shape of the solution is presented in Fig.13 for the case when C_1 is the stationary amplitude of the rolls, and C_2 and C_3 are initially much smaller than C_1. As a result of instability, the perturbation C_2 is growing and reaches a value comparable with C_1 after which the value of C_1 starts to decrease. When C_2 instead of C_1 becomes a solution with stationary rolls, C_3 becomes in relation to it as C_2 was earlier to C_1. But because C_3 was decreasing while C_2 was increasing, C_3 will reach the value of C_2 after a longer time. Therefore, C_1, C_2, C_3 will cyclically replace each other but the periods of this replacement will grow without limit. Such type of the solution is a consequence of the assumption that the perturbations are present only in the initial moment and do not appear in the process of the flow evolution. In the experimental reality, the always existing noise would not allow the damping of the amplitudes to arbitrary small values which would lead to a stochastical behaviour of the cycle periods (see Busse & Clever, 1979).

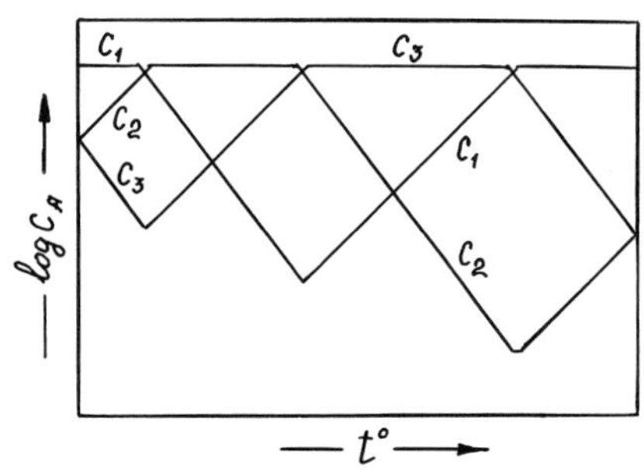

Figure 13. The time dependence of nonstationary solution. (After Busse & Clever, 1979b, courtesy of Springer-Verlag).

For sufficiently large supercritical values of the Rayleigh number, one should use other mathematical methods for studies of the convective motions. Clever and Busse (1979) used a Galerkin method for the case of the layer with two rigid boundaries. As in the case of no rotation, they found stationary solutions and studied their stability (see Fig.1 and their work (Busse & Clever1979)). With rotation, they did it for some small values of the Taylor number $Ta \leq 900$ and for two values of the Prandtl number: 7 (water) and 0.71 (air). In Fig.14 for $Pr=7$ the regions of the roll stability are presented for $Ta = 400$. For larger Ta besides a natural shift into the region of the larger wave numbers with the increase of the rotation, a sharp decrease of the varicose instability boundary takes place, which is replaced by the Kupper-Lortz instability. The latter one becomes the sole instability starting with $Ra > 4050$ for $Ta = 900$. The stable rolls for this value of the Taylor number exist only for $Pr > 1$. For $Ta > 2900$ the computations by Clever & Busse (1979b) confirm the results by Kupper (1970) that stationary rolls are unstable for all values of the Prandtl number.

Experimental studies of the spatial structures in the convective motions of the rotating fluids at small supercritical values of the Rayleigh number were carried out by Krishnamurti (1971), Koschmieder (1968) and T. Rossby (1969). Visual observations by Krishnamurti for water ($Pr = 7$) and silicon oil ($Pr = 100$) showed that at increasing Rayleigh number, two-dimensional rolls are formed on which at a further increase of Ra the lateral instability is arising. Then a stationary state is arising in which perturbations set on with the wave vector of 58° to the roll axis, as the theory by Kupper & Lortz (1969) predicts. The internal structure of the patterns in some regions of the state was changing with a period about minute. The value of the Taylor number at which this state arises for the Rayleigh numbers close to critical ones are found to be lower than the theory predicts and for $Pr = 6.7$ is equal to $Ta \approx 150$ and for $Pr = 100$ is equal to $Ta \approx 700$.

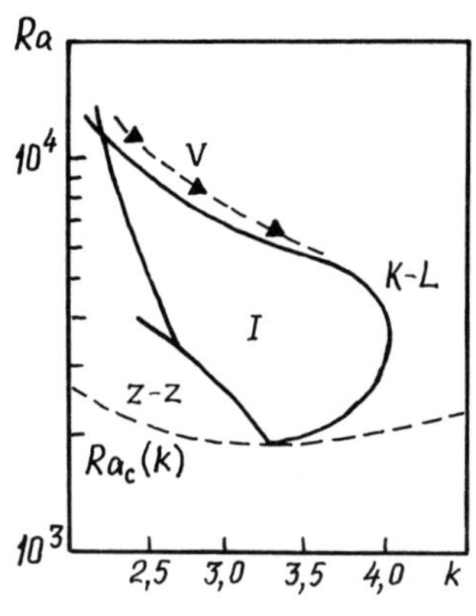

Figure 14: The regions of stability I of rolls for $Pr = 7$ and $Ta = 400$. Boundaries of instabilities: V = skewed varicose, z-z = zig-zag, K-L - Kupperts-Lortz. (after Busse & Clever 1979a, courtesy of Cambridge University Press).

Koschmieder (1968) considered the influence of the rotation on the formation of the concentric rolls in an annulus. For the silicon oil with viscosity 100 cCt (for the comparison, in water the viscosity is four orders of magnitude smaller) this author noted the influence of the rotation starting with $Ta = 8.75$ ($\Omega = 157\,s^{-1}$). In more details this and related works will be considered below in section 2.6, devoted to convective rings.

T. Rossby (1969) also worked with silicon oil but of viscosity 10 cCt and Pr=100 for a range of the Rayleigh number $10^3 - 10^5$ and the Taylor number 0 - $5 \cdot 10^4$. The results are presented in Fig.15 as isolines of the non-dimensional characteristic roll scale $a_{cr} = \frac{l}{h}$ on the plane of $Ra - Ta$. At small rotation rates, Ta <100, the scale does not depend on the rotation rate Ω and is the same as for the case Ω =0. The rolls are observed for $Ra = 2.2 \cdot 10^3$ to $8.7 \cdot 10^4$. In the range of $Ra = 1.1 \cdot 10^4$ to $2.6 \cdot 10^4$, the rolls cease to be two-dimensional and periodical triangular structures are formed on the rolls which are preserved up to Ra of order 10^5.

At larger Ω, the roll structure changes and they become sensitive to the rotation rate and tend to form radially symmetrical structures which might be related to the action of the centrifugal forces. The characteristic scale a_{cr} is decreasing while Ta is increasing. However it is difficult to find any particular dependence in a_{cr} as a function of Ra and Ta.

2.5 Spatial structure at large Rayleigh numbers

Any simple theoretical consideration of convective motions is possible only near the critical stability curve. The linear analysis of the convective equations carried out in section 2.2 gives the dependence of the characteristic scale both for the stationary and oscillatory convection as $a \propto Ta^{\frac{1}{6}}$ and shows independence of the Rayleigh number. At large distances from the stability curve, the flow becomes instationary and irregular. In this case approximate methods are used, some of which are based on the approximation of the so-called "mean" fields. Others use variational principles of the turbulence theory.

The variational theory of turbulent convection was proposed by Howard (1963) and is used for finding the upper limit of the heat flux. In this theory, the heat flux is maximized relatively to integral relationships derived from the equations of motion and continuity. The generalization of this theory for the case of rotating convective fluids was carried out in the works by Chan (1974), Hunter & Riahi (1975), Riahi (1977). According to results of these works, three different regions are delineated on the parameter plane $Ra - Ta$. First, the region of weak rotation $Ta \leq O(Ra)$ in which the rotation is not significant and for the layer with free boundaries there is a solution with a single horizontal mode $a = O\left(Ra^{\frac{1}{4}}\right)$. Second, the region of moderate rotation, $O(Ra) \leq Ta \leq O\left(Ra^{\frac{4}{3}}\right)$, in which the solution has a multimode structure, depending on the rotation rate. However, for any given values of Ra and Ta only a finite number of modes exist here. The number of modes is chosen from the condition of reaching the inequality $(Nu-1)_N \geq O(a_N)$, where N is the number of the mode and a_N is its spatial non-dimensional scale. This condition is always possible for a finite N which increases with the Taylor number. But when this condition is reached, the further increase of the number of the modes is impossible. In the third region of the strong rotation, $O\left(Ra^{\frac{4}{3}}\right) \leq Ta \leq O\left(Ra^{\frac{3}{2}}\right)$, the condition $(Nu-1)_N \geq O(a_N)$ can never be reached which means the presence of an infinite number of wave numbers . For still stronger rotation, $Ta \geq O\left(Ra^{\frac{3}{2}}\right)$, the convective heat transfer is impossible as in the linear theory, because here we traverse the critical stability curve $Ra_{cr} \propto Ta^{\frac{2}{3}}$.

In addition to the considered modes (so-called R-modes) in the case of rigid boundaries (see Riahi, 1977), new modes appear related to the Ekman layer (T-modes). However the existence of the three regions and qualitative characteristics of the modes do not change in comparison with the just described case of free boundaries.

The first experimental studies of the spatial structure of convection in rotating fluids were performed by Nakagawa & Frenzen (1955). They determined the critical curves of the convection onset (see section 2.3) together with the characteristic scale of the spatial structures at three values of the Taylor number near the critical curve. The measured scales confirmed the linear theory results $a \propto Ta^{\frac{1}{6}}$ (see Eq.(2.20)), i.e. the size of the patterns does not depend on the heat flux and is proportional $\Omega^{\frac{2}{3}}$. The first experiments of turbulent convection were carried out by Dikarev (1983). He found that in the region far away from the critical curve, the scale depends not only on the rotation rate but also on the heat flux f, or on the Rayleigh number.

Systematic studies of the convection spatial scale in the range of the flux Rayleigh numbers from 10^6 to $2 \cdot 10^{11}$ and Taylor numbers from 10^6 to 10^{12} were performed by the authors of this monograph (see Boubnov & Golitsyn, 1986) for the plane horizontal layer of water heated from below and cooled from above mainly due to evaporation and conduction from the free surface though some of the experiments were conducted with a rigid upper surface. Much of the material in this book will be devoted to the results of our systematic experiments of the convection. The description of these experimental setting and main procedures of experiments see Boubnov & Golitsyn (1986,1991) and Golitsyn & Grachov (1986).

The main stable stationary state of convection not far above the stability curve in the region of the Rayleigh and Taylor numbers cited above is the convective vortex grid. The system reaches this state after about an hour with an evolution through transient states described in the next section 2.6. The plan and oblique views of the vortex grid are presented in the photographs of Fig.15. The visualization of the flow patterns was done by using bromthymole blue dye whose powder was spread over the upper surface. It solutes rather slow in water and first it dyes the upper colder film of the water and traces the motions of the colder fluid within the vessel. The motion is concentrating in intensive vortex sinks with axis parallel to the rotation axis of the vessel. The sinks are located at vertices of equisided triangles forming a triangle grid. Something similar to this structure was theoretically obtained by Veronis (1959) (see Fig.12) but this structure is different and more complicated.

In our experiments, the basic stable stationary structure is a vortical convective cylinder. The sinking motion takes place along the axis of the cylinder and rising motion is concentrated at the side surface of the cylinder. The relative motion of a fluid particle on the cylinder is a rotation in the same direction as the basic rotation, i.e. the cyclonic motion. At the same time according to Veronis, a fluid particle should move along conical surfaces and in the upper part of it the particles spiral in the same direction as the basic rotation while in the lower part the sign of spiraling is reversing. As we said, we occasionally observed such structures for Ra and Ta near the critical stability curve but they were unstable and soon transformed into the cylindrical forms.

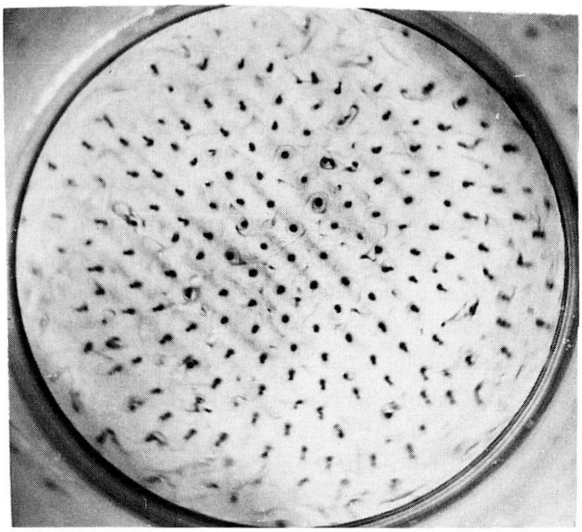

(a)

(b)

Figure 15. Convective vortex grid: (a) plane view, (b) inclined view; $Ra_f = 7.1 \cdot 10^7$, $Ta = 5.9 \cdot 10^8$. (After Boubnov & Golitsyn, 1986, courtesy of Cambridge University Press).

Seems impossible to obtain an analytical description of the observed cylindrical structures in a layer of finite depth, though their radial structure resembles the Sullivan vortices. In the solution found by Sullivan, there is a narrow core with intense sinking motions. The core is surrounded by an external cylinder along which the fluid is ascending. Goncharov & Gryanic (1986) showed that a two-dimensional grid of Sullivan vortices is stable at the rotation as a whole and the distance between the vortices is proportional to $\Omega^{\frac{1}{2}}$ which means that the number density of these vortices is proportional to Ω, the system rotation rate, as in the case of quantum vortices in liquid helium.

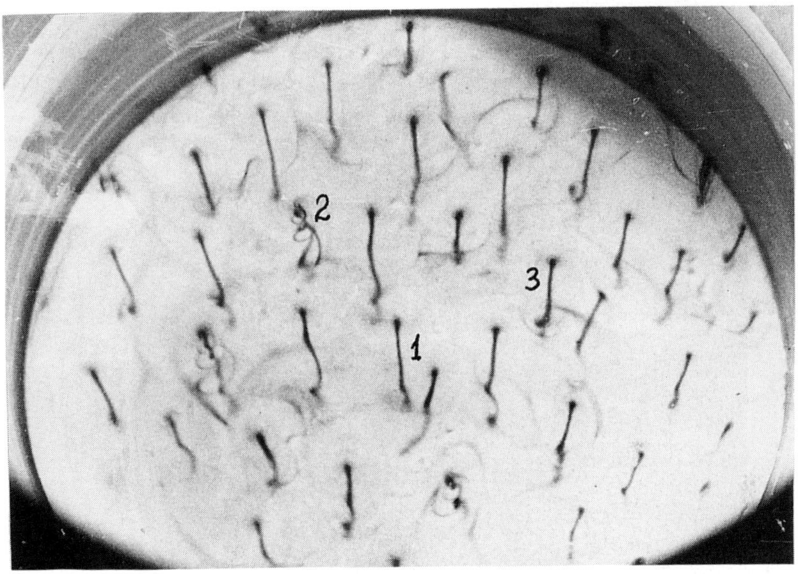

Figure 16: Vortex interactions in a convective grid: 1, interaction of two vortices; 2, formation of a double helix; 3, coalescence of the vortices; $Ra_f = 1.05\cdot 10^8$, $Ta = 5.9 \cdot 10^8$. (After Boubnov & Golitsyn, 1986, courtesy of Cambridge University Press).

If, on the plane $Ra - Ta$,we will move the stability curve either by slowing down the rotation and keeping the same forcing, i.e. at the same Rayleigh number, or increase the forcing while keeping the rotation rate, the convective vortex grid would become less stable. The cylinder would not reach the bottom, the vortices would start to interact with each other. In Fig.16, three stages of this pair interaction are presented. First two vortices start to rotate around each other (as in classic Helmholtz solution for two vortex lines), then the vortices are braided forming a kind of double helix, then the double helix contracts in its diameter and then two vortices coalesce into one of the large intensity. This large vortex weakens with time and near it a new vortex starts to grow and the interaction may continue. Sometimes, a three vortex interaction can be seen but the pair interactions are most frequent. The time period of the interaction increases with the fluid depth and decreases with an increase of the flux f coming through the layer. This time in our experiments was observed to be in the range from 10 to 140 seconds. It was not possible at the time to establish better quantitative observations.

For $Ra = const$ and a further decrease of rotation Ω, irregular motions started to be observed with vortices of various size and intensity moving in random trajectories. The irregular vortices do not have a distinct central sink as vortices in the regular regime (Fig.17). Most of the vortices rotate at the cyclonic direction, but sometimes separate vortices with anticyclonic rotation occur.

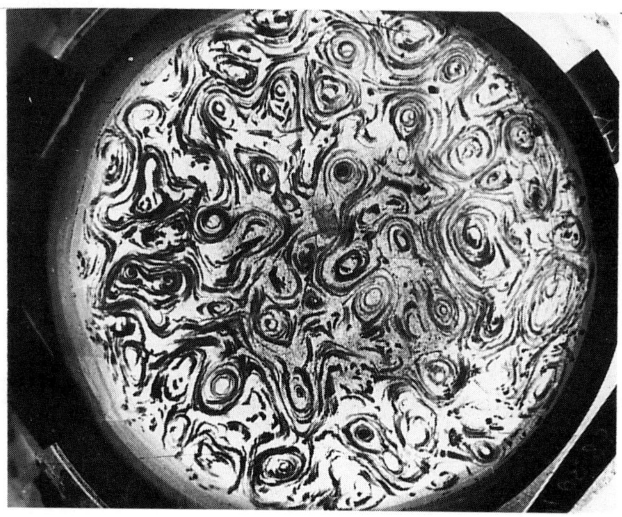

Figure 17: An example of irregular convective vortex pattern; $Ra_f = 2.0 \cdot 10^8$, $Ta = 7.1 \cdot 10^7$.(After Boubnov & Golitsyn, 1986, courtesy of Cambridge University Press).

The transition from fully regular to totally irregular regimes is rather diffusive and occupies about half an order of magnitude in the values of the Taylor number at constant forcing ($Ra = const$). On the plane $Ra_f - Ta$, the transition region goes parallel to the stability curve $Ra_{cr} \propto Ta^{\frac{2}{3}}$ (see Fig.18), but the coefficient of proportionality depends on the value of the heat flux f. The measurements were performed at 4 values of the heat flux, from 85 Wm^{-2} up to 1650 Wm^{-2}. The temperature of the water at these widely different values of f differ considerably which causes significant differences in the value of the molecular coefficients and hence the Prandtl number.

All four curves can be contracted into one by the dependence

$$Ra_{f,2} = C_1 Ta^{\frac{2}{3}} Pr^{-4} \tag{2.41}$$

where $C_1 = 2.6 \cdot 10^5$ (Fig.19).

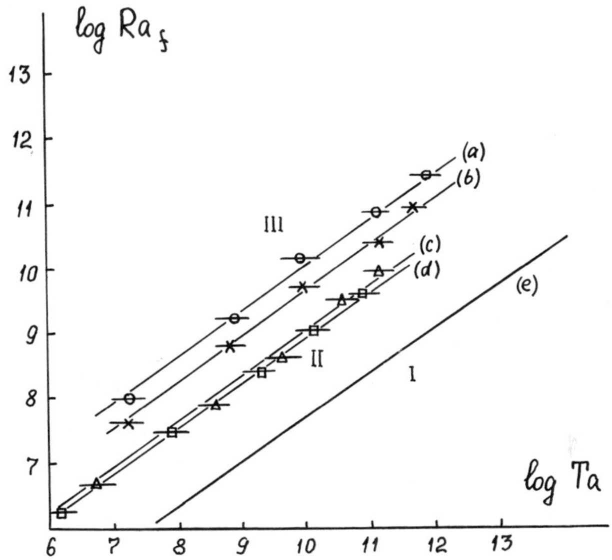

Figure 18: Transition from the regular convective vortex grid to an irregular one for different values of the heat flux: (a) f = 1960W m^{-2}; (b) 690W m^{-2}; (c) 140W m^{-2}; (d) 85W m^{-2}; (e) the neutral stability curve; transition at every point was achieved by decreasing Ω. The right end of a line near each point corresponds to a regular grid, the left end to an irregular one; I refers to the fluid in rigid rotation, II to the regular grid, III to the irregular regime. (After Boubnov & Golitsyn, 1986, courtesy of Cambridge University Press).

The limits of the existence of such a dependence in a wider range of Ra_f and Ta is requires special studies. There is another interpretation of this contradiction procedure, i.e. in terms of the usual Rayleigh number (Boubnov & Golitsyn, 1986). According to this interpretation, the central part of the transition region between the regular and irregular regimes goes parallel to the stability curve $Ra_{cr} \propto Ta^{\frac{2}{3}}$ but with a numerical coefficient about 30 times higher. Evidently, a wider range of Ra and Ta numbers is needed to clarify the situation.

Our experiments give a possibility to determine the characteristic scale of convection at different regimes, simply by counting in the photographs the number of vortices. For the regular regime, when vortices form a triangular grid, the number density of the vortices is related to the distance d between them by the relationship

$$N = \frac{\sqrt{3}}{2d^2} \tag{2.42}$$

If we take as a characteristic scale for d expression (2.19) from the linear stability theory, we obtain

$$N = 0.028\, Ta^{\frac{1}{3}}\, h^{-2} \tag{2.43}$$

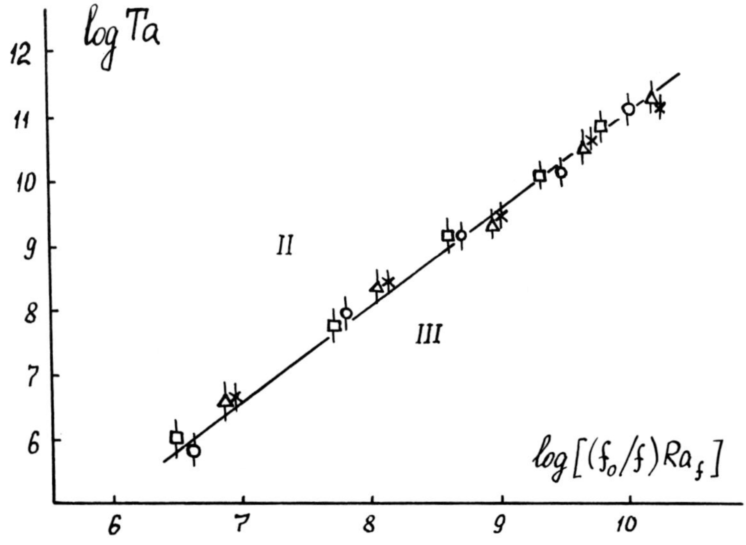

Figure 19: The same as in Fig.18, but the coordinate log Ra_f is replaced by log $Ra_f\left(\frac{f_0}{f}\right)$ where $f_0 = f_3 = 140 W\ m^{-2}$. (After Boubnov & Golitsyn, 1986, courtesy of Cambridge University Press).

The experimentally determined vortex density number agrees well with theoretical dependence (2.43) near the critical stability curve or at small values of the heat flux. At large values of f, or far away from the stability curve, an additional dependence on f is clear. In Fig.20 (in non-dimensional coordinates) we present the results where on the abscissa the ratio of the measured scale to the linear theory one is presented and the ordinate is the logarithm of the ratio $\frac{Ta}{Ra_f}$. It is clear that for $\frac{Ta}{Ra_f} \geq 25$, we have the linear regime and for $\frac{Ta}{Ra_f} \leq 25$ we have another regime which we will call non-linear, though part of the regularity of the flow is still preserved. We note that the ratio $\frac{Ta}{Ra_f}$ can be interpreted as a fourth power of the Kolmogorov internal microscale l_K to the Ekman scale l_E, because $l_K = \left(\frac{\nu^3}{\varepsilon}\right)^{\frac{1}{4}}$, where ε is the rate of the kinetic energy dissipation (see section 1.5), and $l_E = \sqrt{\frac{\nu}{\Omega}}$. We see that how smaller the Kolmogorov microscale and how larger the Ekman scale, the large is the role of the thermal forcing and the smaller the role of rotation.

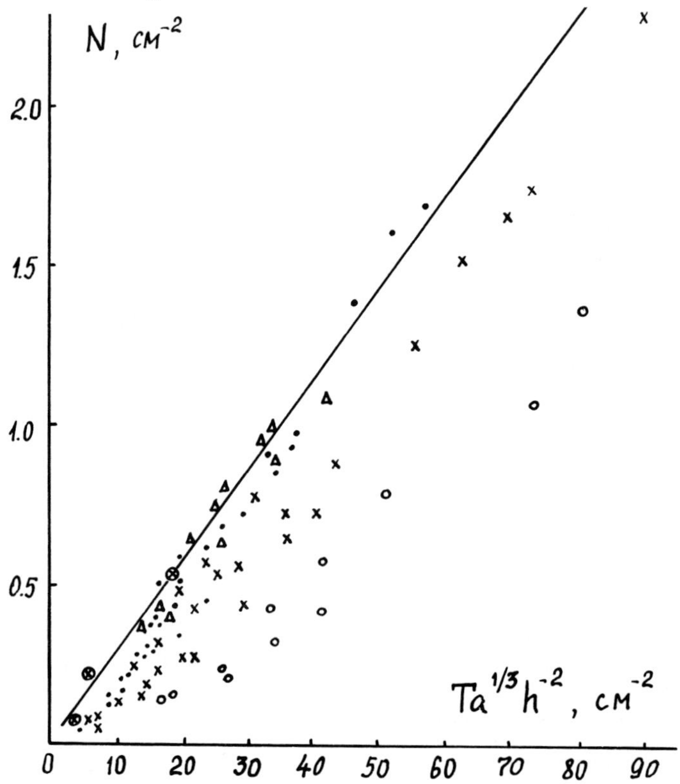

Figure 20: Dependence of the vortex number density N on $Ta^{\frac{1}{3}}\, h^{-2}$; solid line- results of linear theory $N = 0.28 Ta^{\frac{1}{3}}\, h^{-2}$; $\otimes$ points by Nakagawa & Frenzen (1955); $\triangle$, $f = 85 W\, m^{-2}$; $\cdot$, $140 W\, m^{-2}$; $\times$, $690 W\, m^{-2}$; $\bigcirc$, $1960 W\, m^{-2}$. (After Boubnov & Golitsyn, 1986, courtesy of Cambridge University Press).

In the non-linear regime, the number of vortices on a unit area can be approximated by

$$N = C_2 \frac{\Omega}{\nu} Ra_f^{-\frac{1}{6}} = C_2 \, l_E^{-2} \, Ra_f^{-\frac{1}{6}} \qquad (2.44)$$

where $C_2 = 0.033$ and is determined by a linear regression on 104 points. The mean distance between the centers of the adjacent vortices is

$$d = C_3 \, l_E \, Ra_f^{\frac{1}{12}} \qquad (2.45)$$

We may note that the linear dependence of the number of vortices on Ω was also noted in experiments by Hopfinger et al (1982) in which turbulence was generated mechanically by a grid, oscillating in a rotating vessel with water. Evidently, the dependence $N \propto \Omega$ is a common property of strongly mixed rotating fluid.

Numerical calculations for convection in rotating fluids with large Rayleigh number for simulating the motions in the ocean and atmosphere were made by Raasch & Etling (1991), Jones & Marshall (1993), Klinger & Marshall (1994). Raasch & Etling (1991) used so-called a large eddy simulation model and for three values $Ta \approx 10^7 - 10^9$ at $Ra_f \approx 10^7$ they found that the horizontal structure for convective motions was very similar to the laboratory convective structure. The main feature of these structures was a decrease of the horizontal scale of convective cells (updrafts and downdrafts) with increasing rotation. Strong isolated vortices with vertical axes can be observed for increasing rotation, but three points are not enough to check some dependence of the scale on the external parameters. The temperature structure was also similar to experimental one. The main difference with the experiment lay in the selection of the exchange coefficients. In numerical calculations, usually the turbulent exchange coefficients rather than molecular coefficients are used and some difficulties for comparison of laboratory and numerical results may appear. For example, in the paper by Raasch & Etling (1991) the parameters at the regime diagram are in the regular vortex grid region, but the small Prandtl number gives a possibility to shift their results to the region of geostrophic turbulent convection.

Jones & Marshall (1993) and Klinger & Marshall (1994) made calculations for much larger values of the external parameters and checked a dependence for convective structure scale. They also used the turbulent exchange coefficients (different in horizontal and vertical directions) and found that dependence for the vortex density number N also has a weak dependence on heat flux close to one observed by Boubnov & Golitsyn (1986).

2.6 Convective rings

Convective patterns in the form of rings, i.e. motion of a fluid particle along the toruses in rotating fluids was experimentally studied in two different cases: in a shallow fluid at small supercritical Rayleigh numbers and at large Rayleigh numbers, i.e. $Ra > 10^6$. The first case was extensively studied by Koschmieder (1966a, 1966b,

1967, 1968, 1972, 1978, 1986) for non-rotating and rotating fluids. The second one by Boubnov & Golitsyn (1986). In the last case, the pattern is formed as a transient one during spin-up of convective mixed fluid. Here, we consider in some detail the two cases.

For the plane layer at $\Omega = 0$ heated from below and cooled from above in the linear consideration of the problem (see section 2.2) Zierep (1958) found a solution not only in the form of rectilinear rolls, but also a symmetrical one in the form of ring cells or toruses which are stationary and described by

$$w(x,y,z) = A_w\, I_0\left(\frac{ar}{h}\right) F(z) \tag{2.46}$$

where w is the vertical velocity component, A_w the constant, I_0 the Bessel function of zeroth order,and a the wave number. The ring cells have to form at the infinite plane around a point source of disturbance over which the Rayleigh number has the critical value. Experimentally, for the first time the ring cells have been obtained by von Tippelskirch (1959). The influence of rotation on ring cells was studied in the work of Muller (1965) where it was shown that in a rotating fluid a system of concentric ring cells is also arising but projections of the stream lines into the layer boundaries change from the direct radial lines for $\Omega = 0$ to logarithmic spirals in case of rotation. For a rotating fluid layer with free boundaries, the solution for the velocity components has the following form:

$$v_r = A_v\, I_0(ar)\, F(z)$$

$$v_\varphi = A_v\, \Omega^{\frac{1}{2}} \left(\pi^2 + a^2\right) I_0(ar)\, F(z) \tag{2.47}$$

Koschmieder has shown that for slow rotation rates (in his case $\Omega < 0.25$ rev/sec $= 1.57 s^{-1}$) the convective process is developing in the same way as in case of no rotation. At larger values of Ω, the rotation started to be felt. The most characteristic feature of the rotation influence was the formation of ring rolls of alternating larger and smaller sizes which was similar to Koschmieder's (1966) results for a convective layer with the temperature radially alternating at the lower boundary of the layer. In a rotating vessel with diameter of 20 cm at small supercritical Rayleigh number, he could obtain up to ten concentric ring rolls instead of seven at case of $\Omega = 0$ with the other conditions same. But with increasing Rayleigh number, the number of the cells is coming down again to seven which is evidently determined mostly by the aspect ratio $\frac{r}{h}$ with $h = 0.97$. For an increase of the rotation at its moderate rates Ω =3.14 s^{-1}, the origin of the first cell is taking place at the values of the Rayleigh number smaller than the critical one. During an increase of the temperature difference at the boundary, wave-like disturbances appear at the ring rolls which propagate from the outer cell to the vessel center inward. Discussing these experiments, Koschmieder (1978) noted that the main difference between theory and experiment is caused by the presence of centrifugal forces which do not allow the fluid to be in the solid rotation state before the onset of the convection and also by the presence of the side boundaries, as has been already discussed in sections 2.2 and 2.3.

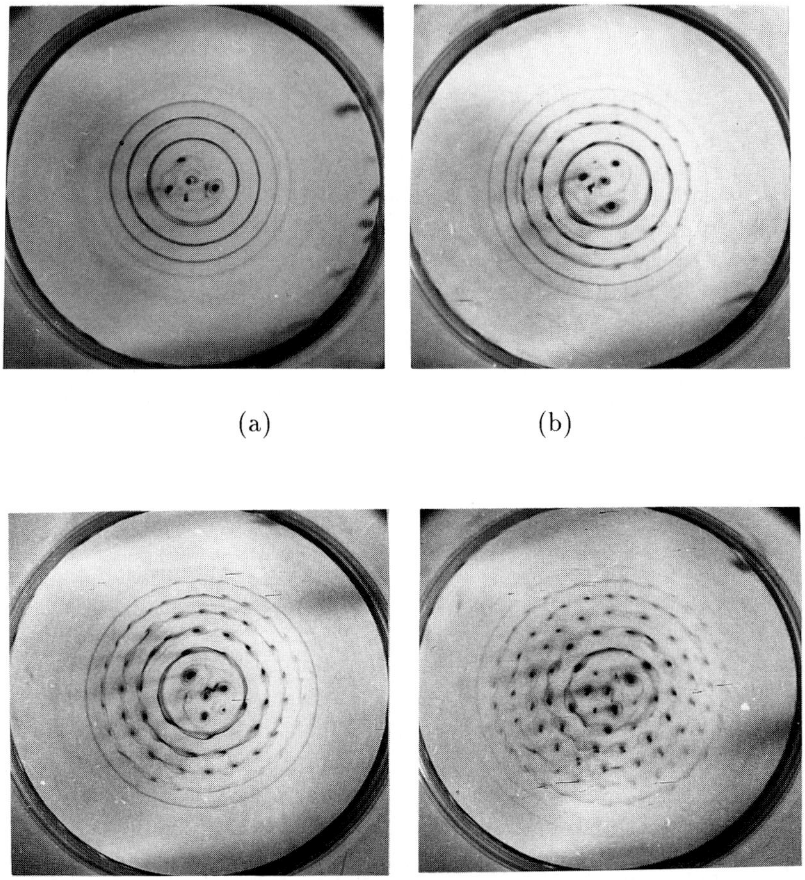

(a) (b) (c) (d)

Figure 21. Ring convective pattern (a) and various stage of its instability (b,c) leading to the transition to the vortex grid pattern (d) at $Ra_f = 3.3 \cdot 10^7$and $Ta = 1.9 \cdot 10^8$. (After Boubnov & Golitsyn, 1986, courtesy of Cambridge University Press).

Externally similar, but different in the mechanisms of origin, a pattern of convective rings is observed at the simultaneous action of convection in a rotating fluid and of a radial gradient of the azimuthal velocity. The existence of such structures in the state of the vessel's spin-up has been noted by Dikarev (1983). With some detail, this question has been considered by Boubnov & Golitsyn (1986). The studies carried out have shown that in a non-rotating fluid the convection exist in a turbulent regime with $Ra \geq 10^6$; then, after some time τ_r counted from the start of rotation concentric ring cells start to form (see Fig.21). These cells start to form at the center of vessel of any form (we checked rectangular vessels, though most our experiments have been performed in a circular cylinder, 21 cm high and 17 cm in diameter). This differs from the shallow cylinder of Koschmieder where the first cell starts to form near the outer wall. We have performed 87 measurements of the first cell formation time τ_r. Our experimental dependence can be approximated in the viscous time units $\frac{h^2}{\nu}$ as

$$\tau_r = 2.06 \frac{h^2}{\nu} Gr^{-\frac{1}{3}} + 20.3 \quad (s) \tag{2.48}$$

where the Grasshoff number $Gr = Ra\, Pr$ and does not depend on Ω, the system rotation rate. We may recall that for the vertical velocity component in turbulent convection in case of no rotation in the laboratory (see e.g. Golitsyn,1980), it was obtained that $w = a_w (\varepsilon z)^{\frac{1}{3}}$, where ε is the rate of the kinetic energy dissipation, $a_w \approx 0.4$ is a numerical coefficient. A similar expression has been obtained for the vertical velocity component in the vortex core; see section 2.12. Then the expression (2.48), disregarding the usually small constant term 20.3 sec, can be presented as a time needed for a fluid particle to pass the layer depth h with an average velocity w:

$$\tau_r \propto \frac{h}{w} \tag{2.49}$$

The convective rings can be observed only at sufficiently fast rotation rates: $\Omega \geq 2\, s^{-1}$. This can be explained by comparing the spin-up time $\tau_{su} = \sqrt{\frac{h^2}{\nu\Omega}} = \sqrt{2}\,\Omega^{-1}\, Ta^{\frac{1}{4}}$ with the characteristic time τ_r in Eq.(2.49). If $\tau_{su} > \tau_r$, the convective rings can not form and the structure of sinking vortices is formed.

The formation of the first ring in the center of a vessel sets a path for future formation of the cells and their evolution. Following the spread of the dye distributed on the layer free surface, we see in Fig.22 that the depth of the dye is about inversely proportional to the distance from the center. A similar picture is observed in the rectangular vessel; the main requirement is the coincidence of the axis of the vessel's symmetry with the rotational axis.

The diameter of the first ring d_m formed in the fluid depends on the rotation rate as Ω^{-1}, the fluid depth and the heat flux. Boubnov & Golitsyn (1986) presented the result of their measurements as

$$d_m = 6.2\, l_E\, Ra^{0.4} \tag{2.50}$$

Figure 22: Side view of the ring convective pattern: the depth of the rings decreases outwards from the center. (After Boubnov & Golitsyn, 1986, courtesy of Cambridge University Press).

The distance between adjacent rings has been observed to change in our experiments from 0.4 to 1.3 cm and it was decreasing from the center to the vessel's rim. The averaged distance r_r could be approximated by the expression

$$r_r = 3.3\, l_K + 0.39\ (cm) \tag{2.51}$$

where $l_K = \left(\frac{\nu^3}{\varepsilon}\right)^{\frac{1}{4}}$ is Kolmogorov's microscale.

The ring structure formed at the fluid spin-up is not stationary. After some time (could be up to 20 min) an instability is formed on the rings (see Fig.21 b,c) caused by changes of the azimuthal velocity component v_φ with the radius. The radial velocity component has opposite signs at the opposite sites of the convergence line visualized by the sinking dye. Therefore, the Coriolis force acts on them at opposite directions causing a shear in v_φ. This sharp shear causes a typical Kelvin-Helmholz instability. The instabilities develop into sinking vortices which in their turn redistribute themselves practically instantaneously into a vortex grid (Fig.21 d). This grid is already the stable formation existing during the experiment if the external conditions do not change. The time of formation of the vortex grid,i.e. the time of existence of the ring structure can be represented after 116 measurements as

$$\tau_v = 101\, \frac{h^2}{\nu}\, Ra_f^{-\frac{2}{3}}\, Ta^{\frac{1}{4}} \tag{2.52}$$

The concentric ring structure as a stationary patterns was obtained by Berdnikov et al (1990) in the presence of the velocity shear at one of the horizontal boundaries. The shear is caused by a slightly different rotation of the horizontal boundary. The case will be considered in section 2.11.

2.7 Thermal structure of regular convective motions

Numerical study of the convective structure of rotating fluids at small supercritical values of the Rayleigh number $Ra \leq 12 \cdot Ra_{cr}$ and small rotation $Ta \leq 10^5$ were conducted by Veronis (1968) and Somerville (1971). For the construction of a numerical solution, they used initially a two-dimensional model with a prescribed wave number a obtained from the linear stability theory. This model has the equation for the stream function and y-component for vorticity (2.30) - (2.32).

The variables are expanded into a series (see section 2.2):

$$\psi = \sum_{m=1}^{M} \sum_{n=1}^{N} a_{mn} \sin(axm\pi) \sin(n\pi z)$$

$$T = \sum_{m=0}^{M} \sum_{n=1}^{N} b_{mn} \cos(a\pi mx) \sin(\pi nz) \tag{2.53}$$

$$u = \sum_{m=0}^{M} \sum_{n=1}^{N} c_{mn} \sin(a\pi mx) \cos(\pi nz)$$

where amplitudes a_{mn}, b_{mn}, c_{mn} are functions of time. The expansion (2.53) is substituted into the equation system (2.30) - (2.32) and after some algebraic transformations a system of ordinary differential equations with respect to time is obtained for variables a_{mn}, b_{mn}, and c_{mn}.

The main feature of the temperature field is the Z-like shape of its mean vertical profile when in the central core of the cell the temperature gradient is positive. This is also typical for the roll convection in case the Ω =0 (see Veronis, 1966). For a slowly rotating fluid, the horizontal distribution of cold and warm flows of the fluid is constrained only by geometry. In the case of strong rotation an additional balance between the horizontal temperature difference and vertical shear of the zonal velocity component is present, the so-called thermal wind which plays an important role in understanding of most problems of motion in rotating fluids. This balance is determined by terms$Pr Ta^{\frac{1}{2}} \frac{\partial u}{\partial z}$ and $Pr Ra \frac{\partial T}{\partial x}$ in Eq.(2.30) and is fulfilled up to the importance of the dissipative and non-linear forces. It is a consequence of the well known geostophic balance in the rotating fluids. The balance is becoming more exact with an increase of the Taylor number and leads to a damping of convective motions. Vertical mean temperature profiles for three values of the Taylor number are presented in Fig.23.

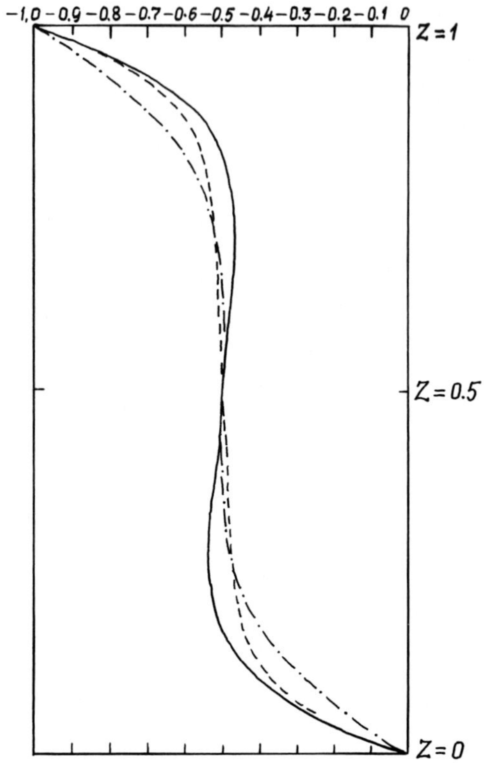

Figure 23: The horizontally averaged temperature, T_m, is plotted against the vertical coordinate for three cases. The strong effect of the constraint for the case $Ta = 10^5$ is reflected in the slope of T_m, which is negative everywhere. In both cases for $Ta = 10^3$, the curves show a mean temperature gradient which is stable at mid-depth. The solid line: $Ra = 1.5 \cdot 10^4$, $Ta = 10^3$, dash-dot line: $Ra = 5 \cdot 10^3$, $Ta = 10^3$, dashed line: $Ra = 7.5 \cdot 10^4$, $Ta = 10^5$. (After Veronis, 1968, courtesy of Cambridge University Press).

A numerical solution has been also obtained by Veronis (1968) for a $Pr= 0.2$ when oscillatory convection is possible together with finite amplitude stationary convection. At slow rotation ($Ta = 10^3, Ra = 1200, a = 0.75$) however, the temperature field does not have the Z-like shape for the vertical mean profile but one can clearly see rather deep intrusions of cold and warm flows from the layer boundaries. With increased rotation ($Ta = 10^4, Ra = 4000, a = 0.85$), the Z-like shape is arising with deep intrusions of cold and warm flows which begin to resemble the mean temperature profile for the non-rotating case. This can be explained from the point of view of the relative role of non-linear and Coriolis terms. The larger the influence of inertia for a decreasing Prandtl number, the smaller is the geostrophy and the thermal wind balance, and hence the smaller is the difference with the case of no rotation.

Later, Somerville (1971) performed a similar study for a more realistic case of two rigid boundaries for $Pr = 6.8$. The characteristic scale of the perturbations a was taken not from linear stability theory but was considered as one of the parameters of the problem. For this case he took into account the results of T. Rossby (1969) that with increasing rotation a substantial decrease of the spatial scale was observed. By varying the scale, Somerville succeeded in obtaining more realistic values of the heat flux through the layer. His temperature field is similar to the one in the work of Veronis (1968) and also changes with increasing rotation.

Experimental studies of the thermal structure at large Rayleigh numbers were carried out by Boubnov (1987), Boubnov & Golitsyn (1990b). In the experimental rotating system a special equipment was installed which allowed a thermocouple to move both in the vertical and horizontal directions. The measurements confirmed that the regular vortex grid consists of vortex cylinders with intense sinks of cold water from the surface along the axis of the cylinder and slow a ascent along its outer surface. Measurements of the mean temperature inside the water layer showed that along the horizontal the mean temperature does not change, but the mean vertical gradient changes substantially with increasing rotation rate.

In the case $\Omega = 0$ at large Rayleigh number, the main temperature decrease takes place in the thermal boundary layers near the layer surfaces. Outside the boundary layers, the mean temperature is constant with height; see curve 1 at Fig.24. With increasing Ω, the temperature vertical profile is declining from the vertical and the angle of declination is increasing with Ω, see curve 2-5 at Fig.24. For a further increase of Ω, the critical stability curve is reached and the vertical temperature profile becomes strictly linear which corresponds to pure molecular conduction.

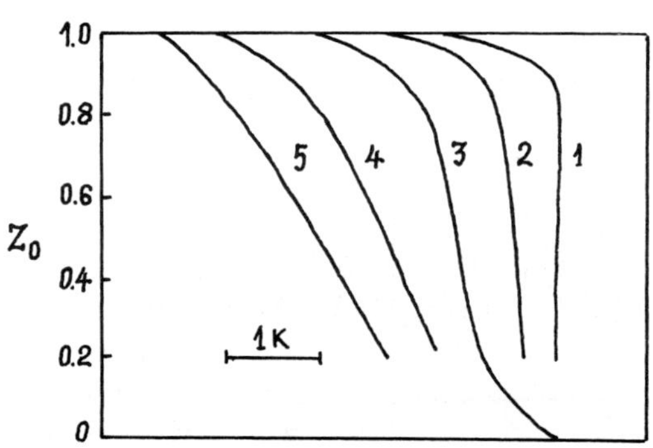

Figure 24: Vertical profiles of the mean temperature for a regular convective grid: $h = 2.7\,cm$, $f = 150 W\,m^{-2}$, $Ra = 2.7 \cdot 10^6$. Curve 1, $\Omega = 0$; 2, $1.04\,s^{-1}$; 3, $2.09\,s^{-1}$; 4, $4.18\,s^{-1}$; 5, $6.28\,s^{-1}$. (After Boubnov & Golitsyn, 1986b, courtesy of Cambridge University Press).

The dependence of the mean temperature gradient on the heat flux through the layer was weak and the main dependence was on the value of $\frac{\Omega}{h}$ with h the layer depth. Using dimensionality considerations, it was possible to represent the mean vertical temperature gradient as

$$\left\langle \frac{dT}{dz} \right\rangle = C_4 \frac{\Omega k}{\alpha g h^2} Ta_f^{\frac{1}{4}} = C_4 \frac{\Delta T}{h} Ra^{-1} Ta^{\frac{1}{2}} Ra_f^{\frac{1}{4}} \tag{2.54}$$

where $C_4 = 4.42$ is a numerical coefficient. One should note that as in the case of two-dimensional rolls, the mean vertical profiles in the convective grid at large Ω is always decreasing with height, at the same time the increase of $\overline{\frac{dT}{dz}}$ is observed with Ω; this can be explained by a stabilizing action of rotation due to the increase of the thermal wind, or geostrophic balance.

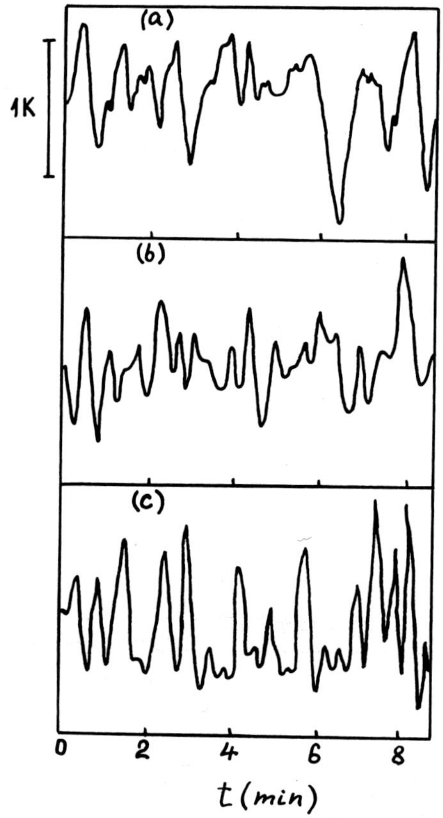

Figure 25: Parts of records of the horizontal time section of temperature taken 1.5 *cm* from the vessel center for the regular regime. The different horizontal positions are: (a) $z_0 = \frac{z}{h} = 0.98$, (b) $z_0 = 0.51$, (c) $z_0 = 0.02$. (After Boubnov & Golitsyn, 1990b, courtesy of Cambridge University Press)

The determination of $\left\langle \frac{dT}{dz} \right\rangle$ has been carried out after 21 measurements for various values of h, f and Ω in the regime of the regular convective grid. Horizontal profiles were also measured for these regimes by moving the thermocouple at various horizons, or by developing a very slow rotation of the main body of the fluid relative to the walls of the vessel and the thermocouple. The effect of the slow relative rotation is reached due to air drag at the surface of the rotating fluid which causes a slow drift of the main fluid mass relative to the walls. Fig.25 presents a horizontal temperature section of the convective vortex grid with the period of the additional rotation of 395 s for the thermocouple at 1.5 cm from the axis of rotation. This is equivalent to the measurements by the thermocouple moving at the circle of 3 cm in diameter with the velocity of 0.024 cm/s. The sections were taken for three horizons $z_0 = \frac{z}{h}$ where z is the distance from the vessel to the thermocouple.

These records allow one to make several conclusions: first, that inside the vortices sufficiently sharp horizontal temperature changes are observed; second, the cylindrical structure of the convective grid is confirmed. Indeed, near the upper surface at $z_0 = 0.98$ the temperature section is observed as the mean temperature and intense pulses of colder fluid corresponding to the sinking motion in the center of the cylinders. At the middle horizon $z_0 = 0.51$, the deviations into warmer and colder sides is about equal. Near the bottom at $z_0 = 0.02$, the ascending motion prevails and the number of positive pulses is about twice the number of minimal ones at $z_0 = 0.98$ which reflects the fact that the colder water sinks at the center and warmer water ascends at the outer wall of the cylinder (which is passed twice by the thermocouple).

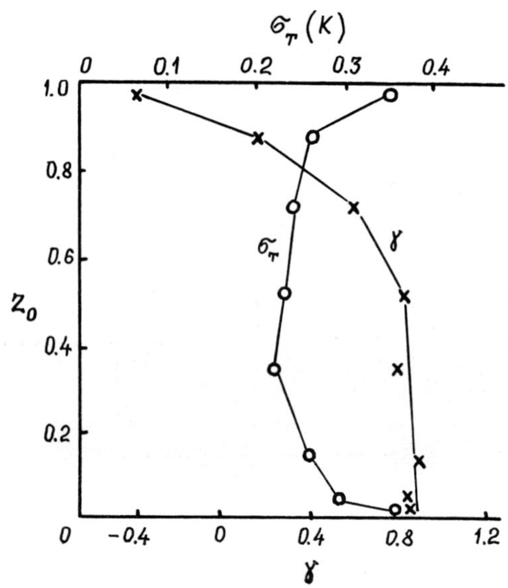

Figure 26: The dependence of temperature variance σ_T and skewness γ_T on z_0 for a regular convective grid. (After Boubnov & Golitsyn, 1990b, courtesy of Cambridge University Press).

The temperature records similar to the ones in Fig.25 were treated for different z_0 to determine the variance σ_T^2 and the third moment $\gamma_T = \left\langle \left(T - \overline{T}\right) \right\rangle \sigma_T^{-3}$. The dependence of these parameters for one of the regimes is presented in Fig.26. The largest variance is observed at the boundary layers and for $0.4 < z_0 < 0.9$ its value decreases with height only weakly; at $0.1 < z_0 < 0.3$ there is a small increase of σ_T. In general one can say that for $0.1 < z_0 < 0.9$ the variance is about constant with height.

The changes in the third temperature moment γ_T reflect the horizontal structure of the temperature field corresponding to motions on the convective cylinders.

2.8 The velocity field in the regular convective regime

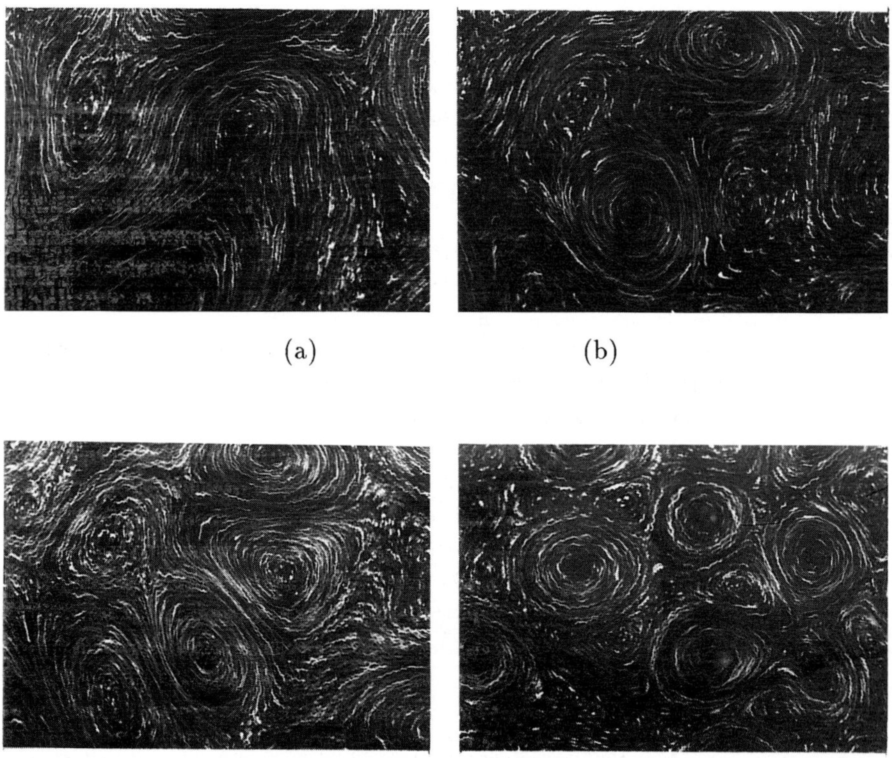

(a) (b) (c) (d)

Figure 27: Horizontal structure of convective motions for different Ω at 2 cm from the upper surface, $h = 10\,cm$, $f = 150\,W\,m^{-2}$, $Ra_f = 5.1 \cdot 10^8$. (a) $\Omega = 0$, (b) 0.52 s^{-1}, (c) 3.14 s^{-1}, (d) 4.19 s^{-1}. (After Boubnov & Golitsyn, 1990b, courtesy of Cambridge University Press).

Theoretically or numerically, the investigation of the velocity fields in convective motions has been carried out alongside with temperature by Veronis (1968), Somerville (1971) and experimentally by Boubnov & Golitsyn (1990b) for certain regimes only. Let us add this velocity field for two-dimensional rolls for large Prandtl number $Pr = 6.8$, following Veronis (1968). The main influence of rotation is in increasing the symmetry of the flow. At small rotation rates ($Ta = 10^3$), the isolines of the zonal velocity are asymmetrical. With the increase of the rotation ($Ta = 10^5$), the isolines are becoming more symmetrical.

Considerably larger is the influence of the increasing Taylor number on the structure of the velocity field in the case of small Prandtl numbers. The almost symmetrical structure of zonal velocity at large Prandtl numbers is replaced by a diagonally asymmetric pattern and with an increase of the rotation rate the motions are intensifying.

Computations by Somerville (1971) at $Pr = 6.8$ for the two rigid boundaries confirmed the results of Veronis (1968) described above with the difference of changing the velocity field near the boundaries at which $v = 0$.

Experimental results of measuring the velocity field in a regular convective grid in water for large Rayleigh numbers were described in Boubnov & Golitsyn (1990b). The maximal vertical velocity in dependence on external parameters was determined by the velocity of the dye spread. The distribution of the horizontal velocity inside the convective cylinder was measured by measuring paths of illuminated aluminum powder particles for a given time; photographs of some convective regimes obtained in this way are presented in Fig.27.

Measurements of the vertical velocity component w inside the convective vortex cylinder were carried out within a large range of external parameters for fluid depth: $h = 5, 10$, and 20 cm, the heat flux $f = 150, 450$, and 1650 Wm^{-2} and $\Omega = 3.14 - 6.28\ s^{-1}$. It was found that the value of w does not depend on the vertical coordinate z, i.e. it is constant inside the layer, depends only very weakly on the rotation rate Ω decreasing at the approach to the critical curve of the convection onset. The main parameters determining the behavior of the vertical velocity are the heat flux f and the fluid depth h. For the non-rotating fluid at large Rayleigh numbers $Ra > 10^6$, the theoretical considerations produce and laboratory experiments confirm the following dependence for the Reynolds number

$$Re_w = \frac{wh}{\nu} = A_w\, Pr^{-\frac{2}{3}}\, Ra_f^{\frac{1}{3}} \tag{2.55}$$

where $A_w = 0.73$ according to measurements. For rotating fluid in the regime of the vortex grid, we have measured 167 values of w inside the vortex cores. The treatment of these data confirm well the expression (2.55) with the correlation coefficient $r = 0.962$ and a slightly larger value of A_w, i.e. 0.80. This means that for similar ranges of the Rayleigh numbers the mean vertical velocities of convection for $\Omega = 0$ and the vertical velocity inside the vortex core in the regular regime are described by the analogous dependencies with close numerical coefficients.

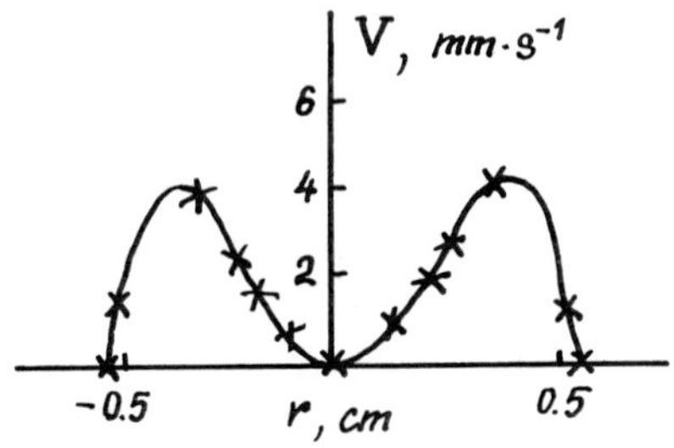

Figure 28: Dependence of the horizontal velocity V on the distance r from the vortex center: $h = 10\,cm, f = 450\,W\,m^{-2}, \Omega = 4.19\,s^{-1}, Ra_f = 2.6 \cdot 10^9$. (After Boubnov & Golitsyn, 1990b, courtesy of Cambridge University Press).

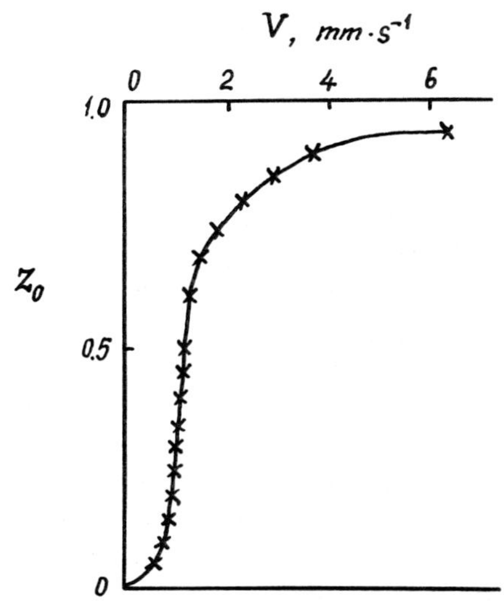

Figure 29: Dependence of the maximal horizontal azimuthal velocity $V_{\max}$ on z_0: $h = 20\,cm, f = 450\,W\,m^{-2}, \Omega = 4.18\,s^{-1}, Ra_f = 4.2 \cdot 10^{10}, Ta = 2.7 \cdot 10^{11}$. (After Boubnov & Golitsyn, 1990b, courtesy of Cambridge University Press).

Measurements of the horizontal velocity field in the convective vortex grid show that the basic motion takes place inside the vortex cylinder and that between the cylinders the fluid, as a rule, has a very small velocity (see Fig.27 d, where the vortex grid is somewhat irregular and the vortices are interacting). The maximal horizontal velocity distribution inside the vortex cylinder at a height of 8 cm, about half way from the bottom is presented in Fig.28 and the dependence of the maximal horizontal velocity on the distance from the bottom is given in Fig.29. These two figures show that the horizontal velocity is maximal on the surface of cylinders, close to zero at the center of the vortex and decreases fast outside the convective vortex cylinder. The dependence of the horizontal velocity on height: the maximal values of the velocity is observed near the surface from which the fluid enters the cylinder and forms the vortex core; for the case of a free surface from which the fluid looses heat due to evaporation, heat conduction and radiation, the instability of the surface cold firm forms the vortex sinks; in the case of two rigid surfaces, the vortex cylinders form at the bottom. The horizontal velocity is decreasing with the distance from the vortex forming surface.

2.9 Turbulent convection in non-rotating fluids

Before considering irregular regimes of convection in rotating fluids, we shall recall some basic regularities of the turbulent convection in non-rotating fluids at large Rayleigh numbers. It is known (see e.g. Krishnamurti,1973) that the transition to turbulence at convection in the plane fluid layer depends very considerably on the value of the Prandtl number: at $Pr \approx 1$, the transition to turbulence takes place at $Ra \approx 10^4$, for water with $Pr \approx 7$ it is at $Ra \approx 10^5$, and for large $Pr \geq 100$ at $Ra = 10^6 - 10^7$. Therefore, one can judge safely that at $Ra > 10^6$ one always observes turbulent convection.

In the turbulent regime of convection, the mean temperature of the fluid does not practically change with the height outside the thermal boundary layers. The temperature fluctuations inside the fluid are determined largely by the processes within the thermal boundary layers. Howard (1963) proposed the following model of the thermal instability development in the layer: originally, a thin boundary layer is formed into which the heat is supplied from the wall by molecular heat conduction, then this boundary layer becomes unstable, as an ordinary fluid layer at passing through the critical Rayleigh number, and is breaking-up by forming thermals; then the supply of heat from the boundary starts again forming the boundary layer and the process repeats itself more or less periodically with time. The characteristic period τ_0 and the spatial scale λ_0 of this process can be presented by external parameters of the system. This has been done and experimentally confirmed by Ginzburg et al (1979), Foster (1969), Foster & Waller (1985):

$$\tau_0 = C_\tau \frac{h^2}{\nu} Ra_f^{-\frac{1}{2}} \tag{2.56}$$

$$\lambda_0 = C_\lambda \, h \, Ra_f^{-\frac{1}{4}} \tag{2.57}$$

The numerical coefficients have been determined by various methods and the results differ within limits of 5% : $C_\tau = 12.7$, $C_\lambda = 48$.

Note that the values of these time and length scales do not depend on the fluid depth and are determined by the heat flux only and molecular constants of the fluid. If we recall that for the developed convection $\varepsilon \approx \frac{\alpha g f}{\rho c_p}$, then the Eqs (2.56) and (2.57), with accuracy up to the Prandtl number, will coincide with the expressions for the Kolmogorov turbulent microscales for time and length: $\tau_K = \left(\frac{\nu}{\varepsilon}\right)^{\frac{1}{2}}$, $l_K = \left(\frac{\nu^3}{\varepsilon}\right)^{\frac{1}{4}}$.

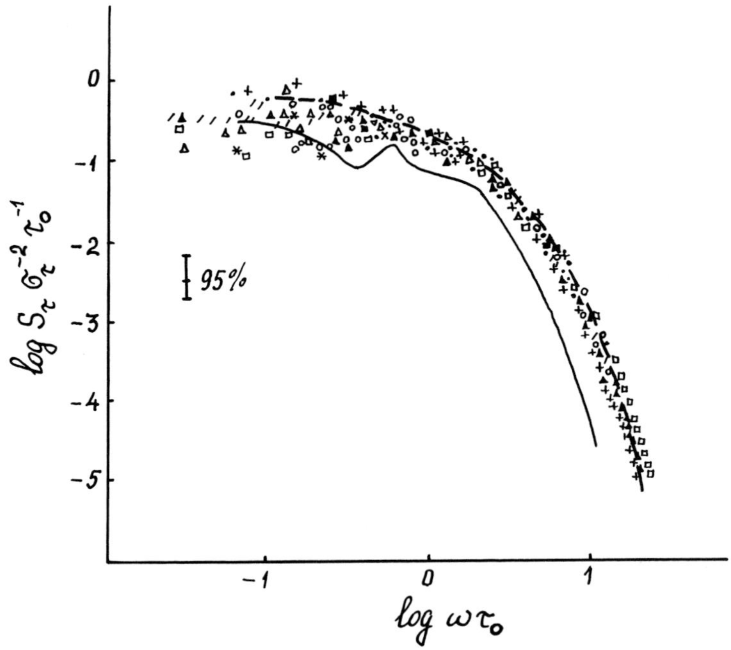

Figure 30. The universal convective temperature spectrum for $\Omega = 0$ in coordinates $S_T\,\tau_0^{-1}\,\sigma_T^{-2}$, $\omega\tau_0$. (After Boubnov & Golitsyn,1990b, courtesy of Cambridge University Press).

Besides measurements of the most probable period in the convective layer τ_0, in many works on free turbulent convection (Zimin & Ketov,1978, Kirdjashkin & Semenov,1983, Deardorff & Willis,1967, Fitzjarrald,1976, Boubnov & Ivanov,1988, Boubnov & Golitsyn,1990b) the temperature fluctuation spectra S_T were also measured in various points of the layer. These studies have shown that within sufficiently large range of the Rayleigh flux numbers $3.2 \cdot 10^7 < Ra_f < 10^{12}$, the temperature time spectrum S_T has a universal form and there are non-dimensional coordinates in which all the spectra coalesce into a single curve. As is shown by Boubnov & Ivanov (1988) such coordinates are $S_t\tau_0^{-1}\sigma_T^{-2}$ and $\omega\tau_0$, where σ_T^2 is the temperature variance and ω is the angular frequency. Fig. 30 represents this patterns as taken from Boubnov & Golitsyn (1990b). Note that this curve represents not only the spectrum of turbulent convection in a plane horizontal layer heated from below and cooled from above, but also the spectrum for the turbulent convection near the heated vertical plane (see Kirdjashkin & Semenov,1983). The specific feature of the turbulent convection in a horizontal layer is the possibility to represent both the temperature variance σ_T^2 and the most probable period τ_0 by the external parameters of the system (for the case of the vertical heated plate, these two parameters σ_T and τ_0 were obtained as a result of statistical treatment of the temperature records and their relationship with external parameters have not yet been found).

The period τ_0 is presented by Eq.(2.56) and the expression for the temperature variance was proposed by Boubnov & Golitsyn (1990b) from dimensional arguments as

$$\sigma_T = C_\sigma \left(\frac{f}{\rho c_p}\right)^{\frac{2}{3}} (\alpha g h)^{-\frac{1}{3}} \tag{2.58}$$

which is a transformation of the expression for the r.m.s. temperature fluctuations over a heated surface obtained first by Oboukhov (1960) by the replacement of the vertical coordinate z by the layer depth h. In Fitzjarrald (1976), Eq.(2.58) was written using non-dimensional system parameters:

$$\frac{\sigma_T}{\Delta T} = C_\sigma \, Nu \, (Nu \, Ra \, Pr)^{-\frac{1}{3}} \tag{2.59}$$

The value of the numerical coefficient C_σ depends on the distance from the lower boundaries z in a rather complicated manner (see Fig.31), reaching a maximum near both boundaries but being approximately constant in the main central part of the layer. For air in the middle of the layer according to the measurements by Fitzjarrald (1976), $C_\sigma = 0.98 \pm 0.05$; however, in Deardorff & Willis (1967) it is equal to $C_\sigma = 1.9 \pm 0.1$. For the water at a distance of 1cm from the heated bottom of the layer 10 cm thick, Boubnov & Golitsyn (1990b) had $C_\sigma = 2.4$. In Fernando et al (1991) for a case of additional heating in the fluid volume, the value of C_σ is constant in the middle of the layer and equal to 0.7 but approaching to the heated bottom, it increases to 2.8. Apparently some other conditions determine the value of the coefficient, may be the aspect ratio. It should be noted that nonstationarity of the heating does not practically influence any results for the turbulent convective regime as for the case of no rotation, as well as for irregular convective regimes in the rotating fluid if the time of the measurements is substantially smaller than the time of a regular change of the mean fluid temperature.

Besides the temperature variance, Fig.31 also presents the third moment, the asymmetry of the temperature probability distribution $\gamma_T = \left\langle T - \overline{T} \right\rangle \sigma_T^{-3}$. It characterizes the ratio of intensities of cold descending and warm ascending fluids. At the lower boundary, thermals of warm water are forming and $\gamma_T < 0$. Near the upper free surface $\gamma_T > 0$ and it is much larger in its module value which reflects the fact that the cold film at the upper surface is thinner than the thermal boundary layer near the bottom and the processes at the film breakdown cause more intensive temperature fluctuations.

One of the features of the spectrum at Fig.30, besides its universality, is an absence of clear linear parts in the bilogarithmic coordinates, i.e. the absence of power dependencies of the spectral density on the frequency. At the same time, in several papers it was proposed to approximate some parts of the spectre by dependencies of $S_T \propto \omega^{\mu_1}$ type with μ_1 = -2, or -5/3, or even -4. However, these approximations describe only short frequency intervals. May be, cases for much larger Rayleigh numbers could reveal more extended intervals for power dependencies.

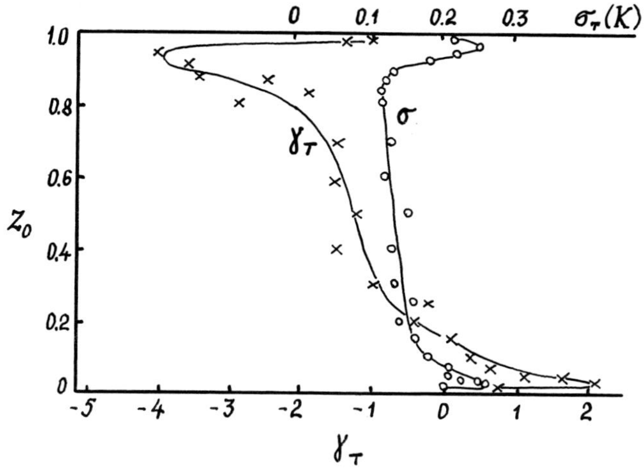

Figure 31: Dependence on $z_0 = z/h$ of the temperature variance σ_T and skewness γ_T, for $\Omega = 0$, $Ra_f = 4.1 \cdot 10^{10}$. (After Boubnov & Golitsyn, 1990b, courtesy of Cambridge University Press).

The characteristics of the velocity fields in the turbulent free convection were considered in Golitsyn (1979,1980) assuming that in this regime the influence of the molecular viscosity is not essential. As a consequence of this assumption, expression (2.55) for the Reynolds number is constructed with the r.m.s. vertical velocity.

The heat transfer in a plane horizontal layer was the main point of attention for many theoretical and experimental works and a detailed review can be found in Goldstein et al (1990). The most popular is the expression for the Nusselt number

$$Nu = C_{Nu} Ra^{\frac{1}{3}} \tag{2.60}$$

It follows immediately from the hypothesis on the independence of the heat flux on the layer depth, which is plausible for the large Rayleigh number. It could be obtained by using the theory of the mixing length from continuity of the heat flux through the thermal boundary layer and from estimates of the velocities at sufficiently large Nusselt number ($Nu >> 1$); see Golitsyn (1979). The regime (2.58) can be considered as an asymptotic one for very large Rayleigh number, but the constant may depend on the Prandtl number. One of the most detailed considerations of the regime was attempted by Grachev (1989) who found for the case of free convection above a heated plate that the coefficient $C_{Nu} = 0.165\,(1 + 0.115\,Pr)^{-\frac{4}{3}}$ for $Pr \geq 0.02$, i.e. for almost all interesting cases. It gives $C_{Nu} = 0.14$ for air and 0.155 for the water in good agreement with experimental data (Katsaros et al,1977, Grachev and Yaroshevich,1989).

For not very large Rayleigh numbers, the theory (Landau and Lifshitz,1989) gives that $Nu = Nu\,(Ra, Pr)$ and numerous experiments give the exponent in (2.58) usually smaller than 1/3.

2.10 Irregular convection regimes in rotating fluids

Here, we return to convection in rotating fluids and consider some characteristics of irregular convective regimes. There are some theoretical studies of these regimes which see as a main goal the heat transfer problem which we will consider in section 2.11. Here, we consider mostly experimental results and their explanations using similarity and dimensional argument following by Boubnov &Golitsyn (1990b), Fernando et al (1989,1991).

2.10.1 Classification of irregular convective regimes

Fig.32, on the coordinate plane $Ra_f - Ta$, presents regions for plane horizontal fluid layer which were studied experimentally. For sufficiently large Rayleigh and Taylor numbers ($> 10^6$), we present the following experimentally based division of the regimes: 1-is the region of no convection and where the heat transfer is purely molecular (in the upper part of this region at small Prandtl number the oscillatory convection is possible); 2-is the region of the regular vortex grid considered in detail in section 2.5 and the critical boundary between region 1 and 2 depends on the nature of the boundary condition (see sections 2.2 and 2.3) but always $Ra_{cr} \propto Ta^{\frac{2}{3}}$. Region 3 and 4 are irregular. The boundary **b** dividing regions 2 and 3 is dividing regular and irregular regimes and is drowns near the center of a fuzzy region of transition as described in section 2.5. The line **c** between regions 3 and 4 is the isoline where the Rossby number $Ro = \frac{U}{2\Omega h}$ is equal to unity (with U the scale of velocity, see Eq.(2.64) below, and h the scale of length). In the region 3, the Rossby number $Ro < 1$ and we will call the region 3 the regime for irregular geostrophic convection and the region 4, where $Ro > 1$ and the influence of rotation is weak, the regime of the usual thermal turbulence.

For determining the Rossby number, we need to know the scale of velocity. It can be determined from the following similarity and dimensional arguments: for the convection, the kinetic energy dissipation or generation per unit mass is equal to $\varepsilon = \frac{\alpha g f}{\rho c_p}$ for large Ra (when $Nu >> 1$) and together with the Coriolis parameter 2Ω and the depth of the layer can be considered as a determining parameter. For sufficiently large scales, at least outside the boundary layers, one can assume it unimportant for all the molecular coefficients entering the governing equations and boundary conditions.

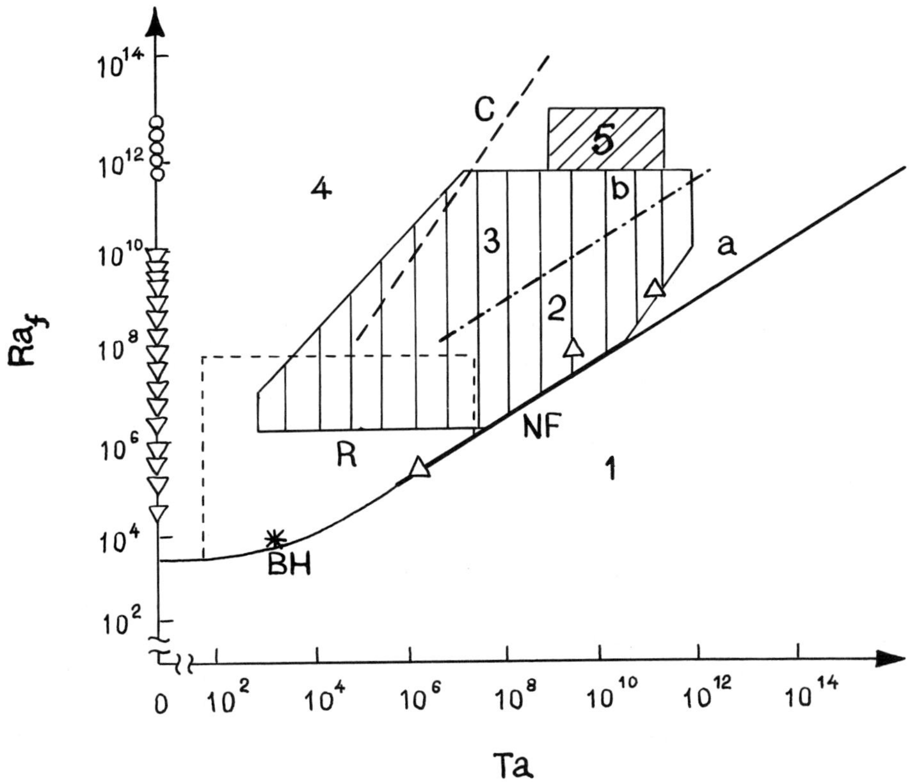

Figure 32: Diagram of different regimes in the plane (Ra_f, Ta) : 1, thermoconductivity regime; 2, regular vortex grid; 3, irregular geostrophic turbulence; 4, thermal turbulence; (a) critical curve for convection onset; (b) curve of transition to irregular regime for *Pr*= 4.5; (c) curve with $Ro = 1$ $(Pr = 4.5)$. Thick solid line, NF, check of stability curve in experiments by Nakagawa & Frenzen (1955); BH, points from Busse & Heikes (1980), R, region studied by Rossby (1969). Region with vertical hatching - experiments Boubnov & Golitsyn (1986,1990b), 5-experiments Fernando et al (1991). (After Boubnov & Golitsyn,1990b, courtesy of Cambridge University Press).

From these three external parameters, one can form a nondimensional number

$$\Pi^* = \frac{\varepsilon}{(2\Omega)^3 h^2} \tag{2.61}$$

The number Π^* can be presented in two forms:

$$\Pi^* = \left(\frac{U}{2\Omega h}\right)^2 \quad , \quad U = \sqrt{\frac{\varepsilon}{2\Omega}} \tag{2.62}$$

$$\text{or } \Pi^* = \left(\frac{l_\omega}{h}\right)^2 \quad , \quad l_\omega = \sqrt{\frac{\varepsilon}{(2\Omega)^3}} \tag{2.63}$$

Taking into account (2.62), we see that we can determine the Rossby number as

$$Ro = \sqrt{\Pi^*} = \frac{U}{2\Omega h} = \sqrt{\frac{\varepsilon}{(2\Omega)^3 h^2}} \tag{2.64}$$

If $Ro >> 1$, than we may assume self-similarity with respect to its precise value which technically means that one of the external parameters entering its definition is not important. Evidently, it is the Coriolis parameter 2Ω. Having been left with only ε and h, we can form the scales of velocity $(\varepsilon h)^{\frac{1}{3}}$ and time $\left(\frac{h^2}{\varepsilon}\right)^{\frac{1}{3}}$ which are characteristic for the thermal turbulence of the Kolmogorov-Oboukhov type. For the case of $Ro << 1$, we may assume a self-similarity with respect to its precise value, i.e. to the precise value of the external length scale h and omit it from consideration. Then, with ε and 2Ω we can form the scales of velocity (2.62) and length (2.63). These arguments were first proposed by Golitsyn (1980), who in a very crude experiment (Golitsyn, 1981) confirmed the expression for the velocity scale (2.62). The same expression was obtained by him from the assumption that the main velocity changes are in the Ekman layers therefore, the main dissipation takes place there also over the scale $l_E = \sqrt{\frac{\nu}{2\Omega}} = h\,Ta^{-\frac{1}{4}}$ and is of order

$$\varepsilon = \frac{\alpha g f}{\rho c_p} \approx \nu \frac{\partial v_i}{\partial x_k}\left(\frac{\partial v_k}{\partial x_i} + \frac{\partial v_i}{\partial x_k}\right) \approx \nu \left(\frac{U}{l_E}\right)^2 \tag{2.65}$$

All the dimensional arguments determine the proper scales up to a coefficient which is usually of order unity. And this the case here too, as it will be described later in this section. If we disregard the condition $Ro = 1$, we obtain the following relationship between the Rayleigh flux and the Taylor number

$$Ra_f = 2^{-2}\, Pr^{\,2}\, Ta^{\frac{3}{2}} = Ra_f^* \tag{2.66}$$

which determines the line **c**, the bordering line between regions 3 and 4 in Fig.32. The condition $Ra_f << Ra_f^*$ will then determine the existence of the geostrophic convection. The reverse strong inequality determines the condition for purely thermal turbulence $Ra_f >> Ra_f^*$.

2.10.2 The thermal structure

The mean temperature within the layer of rotating fluid outside the boundary layers does not change with height for irregular regimes, differing from the case of regular convective vortex grids. Therefore, we shall mainly pay attention to description of fluctuation characteristics of the temperature field which are also easy to measure. An expression for the temperature variance is again easy to obtain from the similarity and dimensional arguments assuming an independence of the variance on height and on the molecular kinetic coefficients. Then the external parameters left are the Coriolis parameter 2Ω, the buoyancy αg and the kinematic heat flux $f' = \frac{f}{\rho c_p}$. From these assumptions we immediately obtain

$$\sigma_{T\Omega}^2 = C_{\sigma\Omega} \frac{2\Omega f}{\alpha g \rho c_p} \tag{2.67}$$

The same relationship can be obtained if we remember the determination for the heat flux (Monin and Yaglom, 1965, 1979)

$$\frac{f_i}{\rho c_p} = C_{iT} \, \sigma_T \, \sigma_{ui} \tag{2.68}$$

and will take the velocity scale (2.62) as the r.m.s. variance σ_{ui} for the velocity. The thermal wind equation obtained from the basic equation (1.5) by taking the curl of its geostrophic approximations reads:

$$2\Omega \frac{\partial \overline{V}}{\partial z} = \alpha \left[\nabla T \times \overline{g} \right] \tag{2.69}$$

This equation together with the velocity scale U and the assumptions that we differentiate here both the horizontal velocity $\overline{V}$ and the temperature over the same length scale, will again produce (2.67).

Our experimental tests of Eq.(2.67) (see Boubnov & Golitsyn, 1990b) have shown that for 30 points measured at 1cm above the bottom for various heat fluxes and depths in the region of the geostrophic convection, the correlation between actually measured values and those calculated by Eq.(2.67) is $r = 0.984$ and the value of the coefficient $C_{\sigma\Omega} = 0.21$.

As for the case of convection without rotation, the universal form for the geostrophic region Eq.(2.67) does not reflect a dependence of the r.m.s. temperature fluctuations on the distance from the layer boundaries. The corresponding study was conducted in a paper of Fernando et al (1991) in which it was proposed to take σ_T from Eq.(2.67) as the temperature scale and as the length scale

$$l_\omega = 4.5 \cdot \sqrt{\frac{q}{\Omega^3}} \tag{2.70}$$

This scale corresponds to the one from (2.63) if we remember that for the free stationary convection the buoyancy flux $q = \frac{\alpha g f}{\rho c_p}$ corresponds to the kinetic energy dissipation rate ε. Coefficient 4.5 in (2.70) is taken from considerations of the section 2.10.4 below.

Spectral functions in the region 4 in Fig.32 does not differ from the spectra described in the preceding section 2.9 for $\Omega = 0$ and also can be described by the universal spectrum of the temperature fluctuations moved toward lower frequencies. However, this movement ends when we reach the region of irregular geostrophic convection as it is shown in Fig. 33, where the spectrum is an average of 26 individual spectra . At these movements in the plane (Ra, Ta), the shape of the spectrum practically does not change. The spectra 1 and 2 in Fig.33 will coincide if we change the time scales τ_0 in the region of irregular geostrophic convection is about three times larger than for the usual thermal turbulence in the plane convective layer. Fernando et al (1991), at somewhat larger values of the Rayleigh number, obtained the temperature spectra similar to ours but they wished to mark a small interval where the spectrum is close to $\omega^{-\frac{5}{3}}$ dependence.

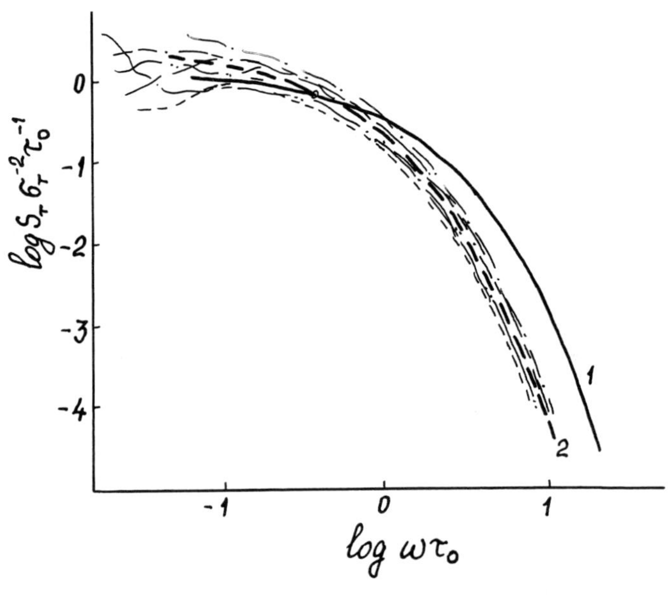

Figure 33: Spectra of $S_T\,\sigma_T\,\tau_0^{-1}$ for different Ω. Curve 1, $\Omega = 0$ (Fig.30); curve 2, mean spectrum for geostrophic turbulence averaged over 26 individual spectra, all of which are within the dashed limits shown. (After Boubnov & Golitsyn,1990b, courtesy of Cambridge University Press).

At a movement to the region 2 of the regular convective vortex grid, the temperature spectrum changes and in the bilogarithmic coordinates a marked linear part appears over about two frequency decades with a slope -2, i.e. $S_T \propto \omega^{-2}$(see Boubnov & Golitsyn,1990b). The nature of this slope is not yet fully understood.

2.10.3 Velocity fields

Studies of the velocity field naturally begin with checking Eq. (2.62) for the velocity scale. This scale is convenient to present in a non-dimensional form as the Reynolds number

$$Re = \frac{V}{h\nu} = C_V \, Pr \, Ra_f^{\frac{1}{2}} \, Ta^{-\frac{1}{4}} \tag{2.71}$$

assuming here V as the r.m.s. of the horizontal velocity component measured in the experiments of Boubnov & Golitsyn (1990b) by the streak photography method. Fig.34 presents results of the test of Eq. (2.71) after 15 experimental points where the ordinate is the experimental value of the Reynolds number and the abscissa is calculated value of $Pr\,Ra_f^{\frac{1}{2}}\,Ta^{-\frac{1}{4}}$. The correlation coefficient between measured and calculated values is $r = 0.985$ and the regression coefficient $C_V = 1.7$. The very first and very crude determination of the coefficient C_w (for the vertical velocity component, see Golitsyn,1980) produced $C_w = 2.3 \pm 0.4$. Fernando et al (1991) found $C_V = 1.70 \pm 0.14$ and this value did not change with sufficient accuracy we may adopt $C_V = 1.7$.

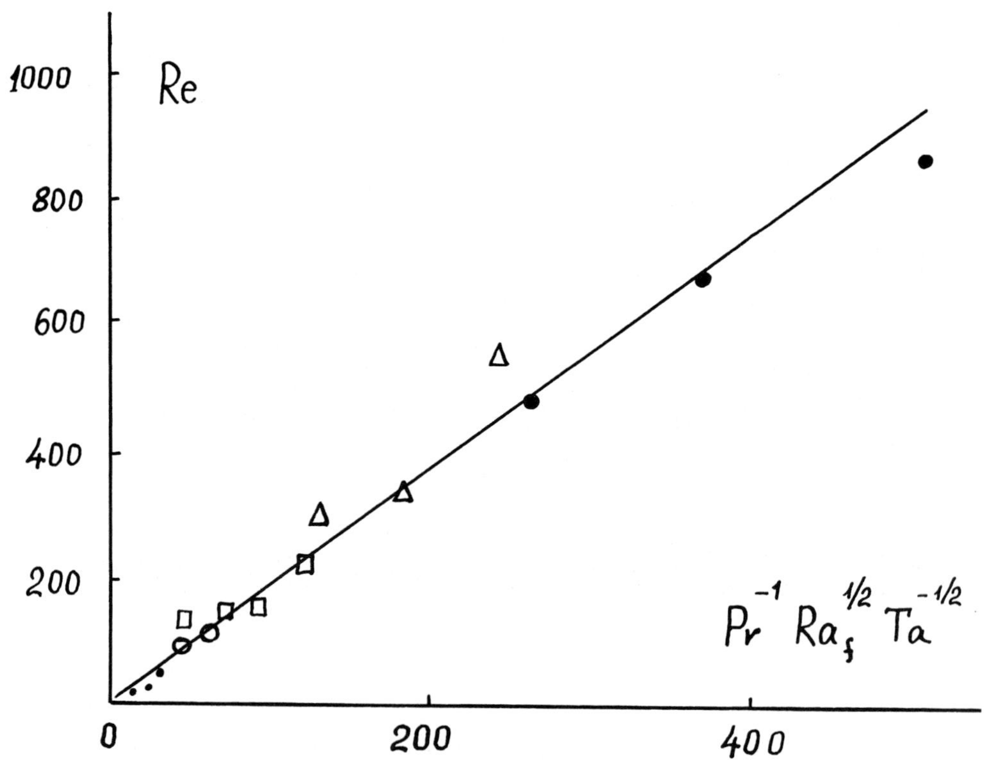

Figure 34: Dependence of the Reynolds number Re on the $PrRa_f^{\frac{1}{2}}Ta^{-\frac{1}{2}}$.

The measurements of vertical velocities is much more difficult in the experiments and besides the measurements in the regular regimes described in section 2.8, these authors did not succeed in overcoming the difficulties by the track photography method. This is because in the vertical plane in the irregular regimes there is no clearly defined vortex motion, as in horizontal plane.

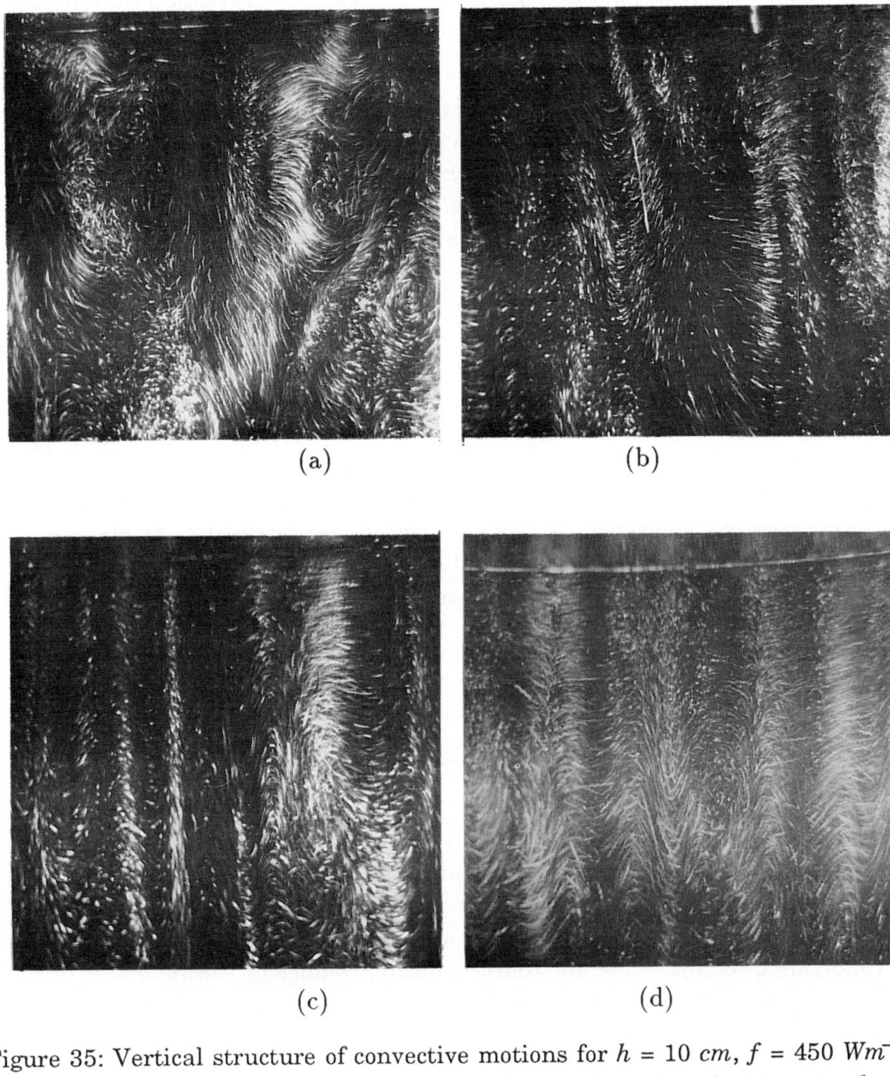

(a) (b)

(c) (d)

Figure 35: Vertical structure of convective motions for h = 10 cm, f = 450 Wm^{-2}, Ra_f = $2.6 \cdot 10^9$ and different values of Ω: (a) $\Omega = 0$, (b) 0.52 s^{-1}, (c) 2.09 s^{-1}, (d) 4.19 s^{-1}. At the upper parts of the photographs one can see the reflections from the fluid upper surface. (After Boubnov & Golitsyn, 1990b, courtesy of Cambridge University Press).

Examples of motions in the vertical plane are shown at Fig. 35 in dependence of the change of the rotation rate. For $\Omega = 0$, we see small-scale motions on the background of the large scale circulation occupying most of the vessel volume. At small values of Ω the motion consists of vortices with arbitrary oriented axes and at large Ω, we see vortices with axes parallel to the rotation axis. This figure illustrates how the increase of the rotation rate translates the regime of convection from irregular three-dimensional motions into the regular vortex grid regime.

2.10.4 The non-stationary convective mixing in rotating fluids

Fernando et al (1991) studied the growth of the convective mixing layer in the rotating fluid from the moment the heat was switched-on. For experimental determination of the depth of this growing mixing layer, two methods have been used: first, by determining the maximum of the r.m.s. horizontal velocity from the track photographs, and second, by thermal measurements of the temperature jump passage through the system of thermistors which correspond to the passage of the thermal turbulence front.

For the fluid at the rest, the growth of the convective boundary layer depth h_b is known to be (see Fernando et al,1991)

$$h_b \simeq 0.3\sqrt{qt^3} \tag{2.72}$$

which follows from the dimensional arguments on the independence of the depth on the molecular kinetic coefficients. It was assumed by Fernando et al (1991) that the growth of the convective boundary layer in the rotating fluid h_Ω would be determined by action of the turbulent eddies on the non-turbulent fluid above the layer. Then the derivative $\frac{dh_\Omega}{dt}$ should be proportional to the horizontal velocity scale for geostrophic convection:

$$\frac{dh_\Omega}{dt} = C_h\sqrt{\frac{q}{\Omega}}, \qquad \text{or} \qquad h_\Omega = C_h\, t\sqrt{\frac{q}{\Omega}} \tag{2.73}$$

Results of testing the relationships (2.73) give $C_h = 0.7$.

2.10.5 Diffusion of passive admixture

The diffusion of a passive scalar in a fluid in the geostrophic convection regime was considered by Boubnov & Golitsyn (1991), but only for horizontal directions. As an admixture, black ink was used placed at the center of the vessel. The change of the area S_D occupied by the ink was registered with time. It was found that the area is linear proportional to time. The horizontal diffusion coefficient can be determined as $D_a = \frac{S_D}{t}$ and this quantity has been measured.

On the other hand, one may assume that the value of the diffusion coefficient D_a should be determined by the velocity scale U from Eq. (2.62) and by the fluid depth h. Therefore

$$D_a = C_D\, k\, Ra_f^{\frac{1}{2}}\, Ta^{-\frac{1}{4}} \tag{2.74}$$

where k is the thermodiffusion of the fluid.

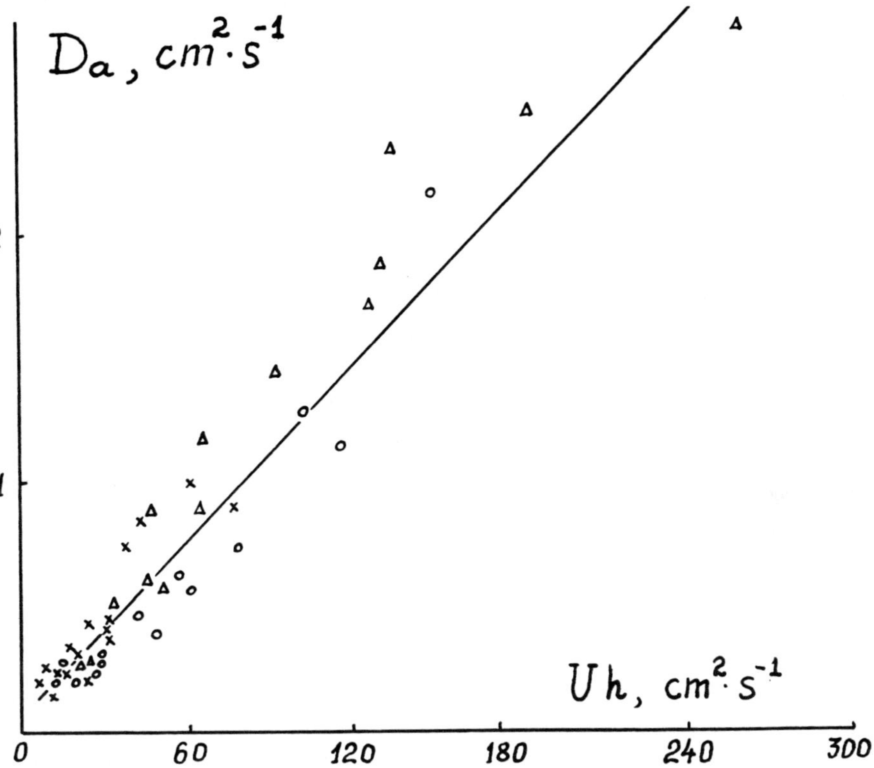

Figure 36: Dependence of the diffusion coefficient D_a on $U\,h$. (After Boubnov & Golitsyn, 1991)

Results of the experimental tests of this formula are presented in Fig.36. For 43 points with the correlation coefficient r =0.93, we have obtained $C_D = 1.25 \cdot 10^{-2}$ with a small additive constant C_{D0} =0.07 cm^2s^{-1} of the order of molecular kinetic coefficients for water. In these experiments, the value of the coefficient C_D was in the range 0.13 to 2.85 cm^2s^{-1}. We may compare these values with characteristic values of molecular diffusion coefficients in liquids which are of order $D_m = 1\,cm^2day^{-1} = 1.2 \cdot 10^{-5}$ cm^2s^{-1} . We see that the diffusion in regimes of geostrophic convection in our experiments were 4-5 orders of magnitude larger compared to the purely molecular one. At the same time, it is smaller than the value of the eddy diffusion coefficient estimated by the Richardson-Oboukhov formula for the case $\Omega = 0$:

$$D_R \approx 0.1\,\varepsilon^{\frac{1}{3}}\,h^{\frac{4}{3}}. \tag{2.75}$$

The ratio of these two coefficients is

$$\frac{D_R}{D_a} \approx 8\,Pr^{\frac{1}{3}}\,Ra^{-\frac{1}{6}}\,Ta^{\frac{1}{4}}. \tag{2.76}$$

2.10.6 Relationship between irregular convective regimes with and without rotation

The difference between irregular regimes in the regions 3 and 4 of Fig.32 lies in the role of the rotation, or, more specifically in the value of the Rossby number *Ro*. In the region 4 of the geostrophic convection the velocity scale is determined by (2.62) and taking into account the numerical coefficient $C_V = 1.7$ from section 2.10, we can write it as

$$Ro = C_V \left(\frac{\alpha g f}{4\Omega^3 \rho c_p h^2} \right)^{\frac{1}{2}}. \tag{2.77}$$

This number gives a possibility to relate some convective characteristics in the non-rotating fluids with the corresponding ones in the rotating fluids by comparing Eqs. (2.58) and (2.62), (2.55) and (2.69), (2.64) and (2.75)

$$Re_V = Re_w\, Ro^{\frac{1}{3}}, \qquad \sigma_{T\Omega} = \sigma_T\, Ro^{-\frac{1}{3}}, \qquad D_a = D_R\, Ro^{\frac{1}{3}}. \tag{2.78}$$

Therefore the characteristics for the geostrophic convection regime can be obtained from the corresponding characteristics of the usual thermal turbulence by multiplication or division by $Ro^{\frac{1}{3}}$, i.e. they all have square root dependence on the rotation rate Ω, since from the Eq.(2.77) $Ro \propto \Omega^{-\frac{3}{2}}$.

2.11 Heat transfer

2.11.1 Theoretical studies

The first theoretical studies of the integral characteristics of convection (among which is the heat flux through the layer) were the next step after investigations of the motion onset and required taking into account non-linear terms. First calculations of the heat flux were carried out at studies of convective motions in structures of various forms: rolls, square and triangular cells. It was found that for the same value of the Rayleigh number for these structures, the Nusselt number is different for these forms. After studies of the stability of the structures, it was shown that the only stable structure are the two-dimensional rolls. Numerical studies by Veronis (1959,1968) of the finite amplitude convection of the rolls were carried out for a plane layer with free boundaries for a range of the Taylor number from 0 to 10^5 and for two values of the Prandtl number: *Pr*= 6.8 which is an order of magnitude higher than the critical Prandtl number *Pr* for the onset of oscillatory convection and *Pr*= 0.2 <*Pr*.

According to these calculations, at *Pr*=6.8 the value of the Nusselt number at the constant ratio $\frac{Ra}{Ra_{cr}}$ is increasing with Ta. At small values of the Prandtl number, the transition from the oscillatory to the finite amplitude convection, or to the monotonic stationary convection causes a sharp change in the value of the heat flux and this transition depends on the wave number chosen for the calculations and determining the motion structure. This wave number was one of the determining parameters, therefore there was no well defined dependence as for heat flux.

Having in mind a strong influence of the boundary conditions on the heat transfer, Clever and Busse (1979) conducted similar studies for a plane layer with rigid boundaries. According to calculations of two-dimensional flows at the fixed Rayleigh number for the Prandtl numbers $Pr > 1$, the heat through the layer always decreased while the Taylor number increased. This contradicted experimental results by T. Rossby (1969); therefore, it was assumed that the difference was caused by three-dimensional effects or by a change of the characteristic wave number of the convective rolls. For the Prandtl numbers lower than the critical one, the Nusselt number naturally reflects changes in the forms of motion. For instance, at $Pr = 0.1$ and 0.36 and $Ra = Ra_{cr}$ there is a finite amplitude convection and then $Nu > 1$ at $Ta = 10^4$, but at $Pr = 1$ and at the same Taylor number, $Nu = 1$. The influence of subcritical finite amplitude convection for the case of rigid boundaries is intensified due to damping of oscillatory instability at these boundaries which is confirmed by the computational results. At the rotation increase up to $Ta = 4 \cdot 10^4$, subcritical convection vanishes and then $Nu = 1$ at $Ra = Ra_{cr}$. For cases of $Pr > 1$, the subcritical convection does not exist at rigid boundaries.

The calculations by Clever and Busse (1979) also revealed a weak dependence of the Nusselt number on the Prandtl number for $Pr > 1$ and a marked increase of the Nusselt number with the decrease of the Prandtl number for $Pr < 1$ (see Fig.37). This differs from the behavior in case of no rotation when the heat flux increases monotonically with the Prandtl number for small supercritical Rayleigh numbers.

Studies of the heat flux at large Rayleigh numbers for $Pr \to \infty$ were carried out by Chan (1974), Hunter & Riahi (1975), Riahi (1977) and their results for the spatial structure of the motions have been described in section 2.6. Optimal values of the Nusselt number were found, i.e. the values when the heat flux through the layer was at its maximum. These values of the heat flux were forming a continuous curve on the plane (Ra, Ta). The plane could be divided into several regions, the convection in which depends on the type of the boundary conditions. For the rigid and free boundaries at small rotation rates $Ta \leq O(Ra)$, the motions is undistinguished from the ones in non-rotating fluids and the values of the Nusselt number are practically the same, as described in section 2.9. At moderate rotation rates for free boundaries $O(Ra) < Ta < O(Ra^{\frac{4}{3}})$, the values of the Nusselt number always decrease with increase of the Taylor number, while in the case of rigid boundaries and $O(Ra) < Ta < O([Ra \log Ra]^{\frac{4}{3}})$, the Nusselt number may increase for fixed Ra with Taylor number increased; the effect had been observed by T.Rossby (1969). This effect is explained by means of the additional influence of the Ekman boundary layers. For fast rotation at the rigid boundaries $O([Ra \log Ra]^{\frac{4}{3}}) < Ta < O(Ra^{\frac{3}{2}})$ and free boundaries at $O(Ra^{\frac{4}{3}}) < Ta < O(Ra^{\frac{3}{2}})$, the expression for the Nusselt number does not depend on the type of boundary conditions and has the form

$$Nu = C_{Nu} Ra^3 Ta^{-2}, \tag{2.79}$$

where $C_{Nu} = 0.0052$ for the rigid boundaries and 0.0032 for the free boundaries. The dependence (2.79), as well as the corresponding expression (2.60) for the heat flux in a non-rotating fluid, reflects the fact that the heat flux passing through the layer does not depend on the layer depth h. It also follows from Eq.(2.79) that isolines Nu = const at large Taylor numbers are parallel to the critical curve $Ra_{cr} \propto Ta^{\frac{2}{3}}$.

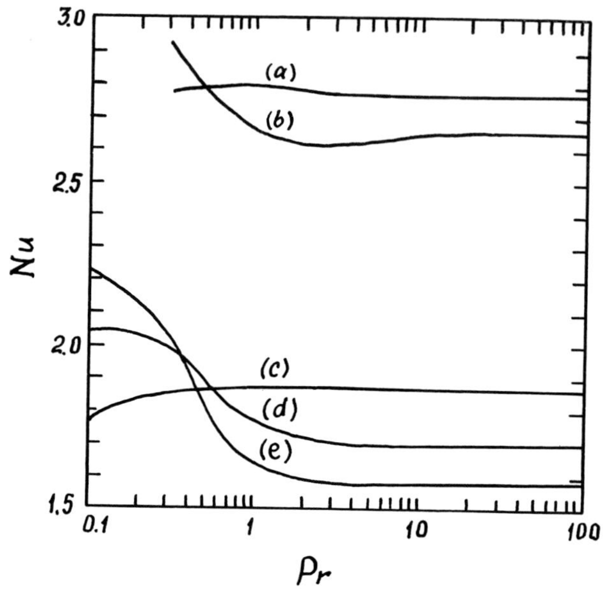

Figure 37: The Nusselt number as a function of the Prandtl number for different values of Ω and $Ra - Ra_{cr}$. $Ra - Ra_{cr} = 10^4$: $(a)\,\Omega = Ta = 0$; $(b)\,Ta = 2.5 \cdot 10^3$. $Ra - Ra_{cr} = 2 \cdot 10^3$: $(c)\,Ta = 0$; $(d)\,Ta = 900$; $(e)\,Ta = 2.5 \cdot 10^3$. (After Clever & Busse, 1979a, courtesy of Cambridge University Press).

2.11.2 Experimental studies

The first experiments with measurements of the heat flux in a rotating fluid were performed by Goroff (1960) for mercury with Pr=0.025. The experiments showed that there was a sharp change in the slope of the curve $Nu = Nu(Ra)$ at the transition from the oscillatory convection to the monotonic one. Very detailed and extensive measurements of the heat fluxes in a plane rotating layer were carried out by T. Rossby (1969) for water with $Pr = 6.8$ and mercury which allowed him to consider both monotonic and oscillatory regimes of convection. For the case of water, his main results are presented in Fig.38, where the isolines $Nu = const$ are presented on the plane (Ra, Ta).

The main features at this figure are the following: (i) at small values of the Taylor number the heat flux through the layer is the same as in non-rotating fluid; (ii) at moderate rotation and at Ra =const for $Nu \geq 3$, the heat flux slightly increases with Ta and there is a maximal value of the Nusselt number at some $Ta = Ta(Ra)$ which corresponds to a minimum on the isolines $Nu = const$. The position of the minimum is determined by the expression $Ra_m \approx 206 \cdot Ta^{0.63}$ and for a fixed Nusselt number the ratio of the Rayleigh number at Ω =0 for the Rayleigh number at this minimum is approximately constant and equal to 0.6. As an explanation of this phenomenon, T. Rossby supposed that there was a change in the structures of the boundary layers and at a given Rayleigh number the maximum of the Nusselt number is observed when the depth of the thermal boundary layer became about equal to the length of the Ekman layer; (iii) isolines of the Nusselt number at large values of Ra and Ta tend to became parallel to the critical curve $C - C$ i.e. the isolines $Ra \propto Ta^{\frac{2}{3}}$. It should also be noted that at these values of Ra and Ta the experimental stability curve of T.Rossby appreciably differs from the theoretical curve by Chandrasekhar (see sections 2.2 and 2.3) though in the space between this curve and the experimental one, the Nusselt number exceeds unity only weakly.

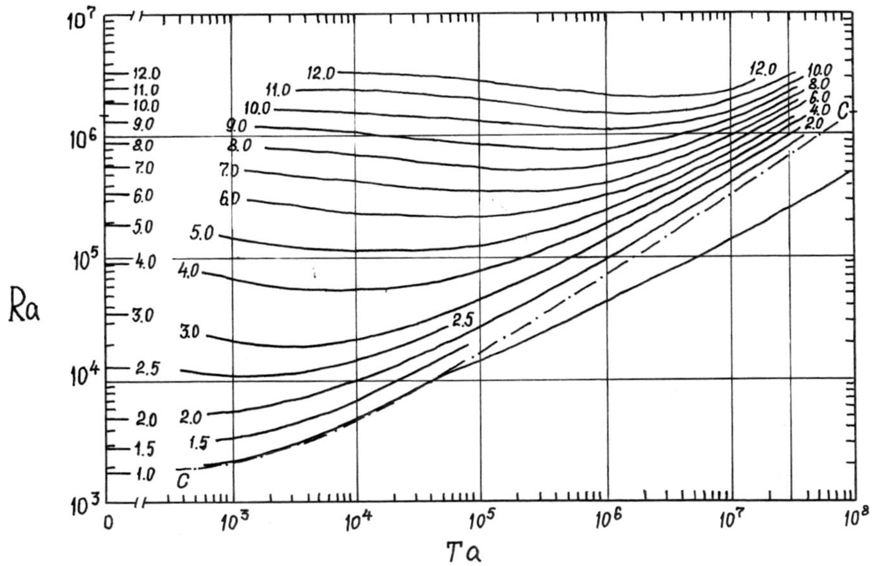

Figure 38: Lines of constant Nusselt numbers as a function of Rayleigh and Taylor numbers in water with Prandtl number Pr = 6.8. Except for small values of the Nusselt number at large Taylor numbers, the error in the Nusselt number is $\leq \pm 2\%$. (After Rossby, 1969, courtesy of Cambridge University Press).

A similar diagram of the isolines Nu = *const* on the plane (Ra, Ta) for the mercury with Pr = 0.025 is shown in Fig.11. Here the line $C - C$ is for monotonic convection and $O - O$ for the oscillatory one and the experimental curve $T - T$ determines the boundary of the transition from regular oscillations to irregular ones. One should note a strong difference of experimental and theoretical results at $Ra <$

$2 \cdot 10^4$ and their good agreement for $Ra > 10^5$. Starting from the last value of the Rayleigh number, the isolines of the Nusselt number tend to be parallel to the critical stability curves, i.e. here we also have the isolines $Ra \propto Ta^{\frac{2}{3}}$.

Interactions between thermal and hydrodynamic characteristics of the motion regimes can be also tracked by an analysis of the dependencies $Nu(Ra)$, i.e. by cross-sections of the diagrams for isolines $Nu = const.$ This question is considered in details by Rossby (1969) for three values of Taylor number and by Lucas et al (1983) and Phontenhauer et al (1984,1987), where the influence of the aspect ratio $\delta = \frac{r}{h}$ was studied for $\delta = 7.81$, 4.93, 3.22 and 1.97 and for Prandtl number varying from 0.49 to 0.76 in dependence of the temperature of the working fluid which was liquid helium-4 (^4He) in these experiments.

The experiments showed that the initial slope for the curve $Nu(Ra)$ depends substantially on the value of δ . For the slope determination, the values of the measured Nusselt number were expanded into a series:

$$Nu - 1 = \sum_i N_i \left(\frac{Ra}{Ra_{cr}} - 1 \right), \qquad i = 1, 2, 3. \tag{2.80}$$

As was shown by Behringer & Ahlers (1982), Charlson & Sani (1975) for the non-rotating fluid, the initial slope increases with δ which corresponds to a stabilizing influence of the side walls on the convective onset. The reverse picture is observed in the rotating fluid (see Fig.39 where the dependence of the coefficient N_1 from (2.80) on Ta is shown for various values of the aspect ratio), where for the range $4 \cdot 10^4 < Ta < 2.5 \cdot 10^5$ the value of N_1 is larger for smaller δ.

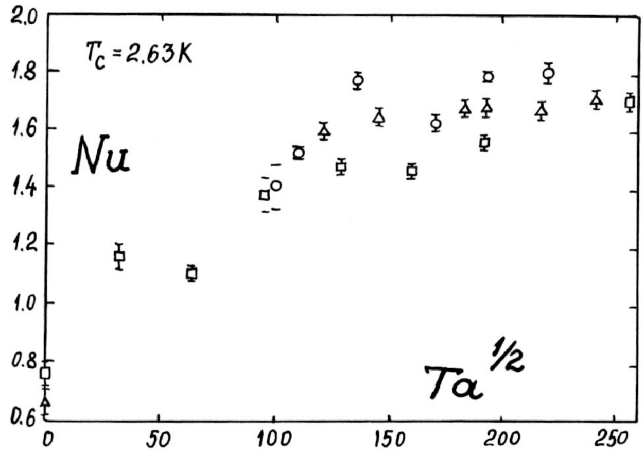

Figure 39: Influence of aspect ratio δ on the initial slopes; Nusselt number Nu is shown as a function of $Ta^{\frac{1}{2}}$ with δ as a parameter: □ – $\delta = 3.91$; △ – $\delta = 2.48$; ○ – $\delta = 1.61$. (After Phontenhauer et al, 1984, courtesy of Cambridge University Press).

Investigations by Lucas et al (1983), Phontenhauer et al (1984,1987) on the dependence $Nu(Ra)$ showed that for the liquid helium ($Pr \approx 0.5$) the Nusselt number is always decreasing with the rotation rate, i.e. as for mercury there are no minima of the isolines $Nu = const$ on the plane (Ra, Ta).

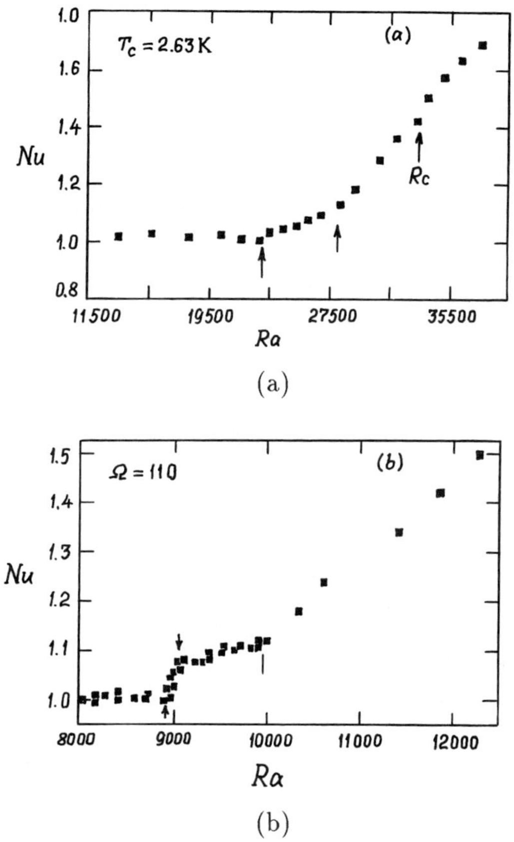

Figure 40: Two different heat-transfer sequences displaying various convection modes for $Ra < Ra_{cr}$: (a) $\delta = 0.99$; (b) $\delta = 1.61$. The arrows mark the assumed transitions from one convection mode to another. The expected values of Ra_{cr} for standard convection are also indicated. (After Phontenhauer et al, 1987, courtesy of Cambridge University Press).

Studies of the heat transfer in the region of subcritical convection showed a marked change of the curve slopes at the replacing of the one convective mode by another and changes of these dependencies for various values of the aspect ratio δ (see Fig.40). For the largest value of $\delta = 3.91$, only one transition was observed at $Ra < Ra_{cr}$, while for smaller values of $\delta = 1.97$ or 3.22 a sequence of transitions was observed . In a non-rotating case, a sequential change was observed in the slope of the curve $Nu = Nu(Ra)$ while at Fig.40 one can see a sharp stepwise jump after which a smooth dependence is observed. It is evident that the sharp change corresponds to a transition to new convective modes at the subcritical convection.

In the experiments by Boubnov and Golitsyn (1990b) estimates were attempted of the heat flux passing through the convective layer in the regime of the vortex grid by measuring the gradient of the mean temperature. Results of the measurements are presented in Fig.26 and approximated by Eq.(2.54). The Nusselt number can be presented as the ratio of the actual "eddy" transfer coefficient to the molecular one:

$$Nu = \frac{k_T}{k}, \qquad k_T = \frac{fh}{\rho c_p \Delta T} \geq \frac{f}{\rho c_p T_z}. \tag{2.81}$$

Because the measured value of the mean temperature gradient $T_z \leq \frac{\Delta T}{h}$ as seen from Fig.24 in the presence of motion, the inequality becomes stronger with movement away from the critical stability curve of the convective onset. The values of T_z in (2.54) and Fig.24 are parts of the thermal boundary layers; therefore, if we put $k_T = \frac{f}{\rho c_p T_z}$, we estimate the Nusselt number from above. Using Eq.(2.54) for T_z, substituting it into (2.81) and remembering that $Ra_f = Ra\, Nu$, we obtain Eq.(2.79) with the coefficient 0.042 which is about an order of magnitude greater than in the theoretical expression (2.79). The question requires direct experimental resolution by measuring the heat flux in the vicinity of the stability curve.

One should note that if (2.79) is valid at some region above the critical stability curve, it would mean that $f \propto (\Delta T^3)\, \Omega^{-4}$, which shows a very strong dependency of the heat flux both on the temperature difference and, especially, on the rotation rate. This may be of some interest to technological applications. In any case, one should expect a strong influence of the rotation on the heat transfer near the stability curve.

2.12 Convection at a non-parallel rotation axis and gravity acceleration

In nature and in technical applications, the vectors of angular rotation and gravity acceleration are usually not parallel to each other. Chandrasekhar (1961) has shown that results of the linear stability theory would not change for the case when there is an angle θ between $\overline{\Omega}$ and $\overline{g}$ if one would consider only the component $\Omega \cos\theta$. Confirmation and expansion of these results were carried out in papers by Flasar & Gierasch (1978), Hathaway et al (1979,1980) which will be considered in Chapter 4 where we consider the presence of both vertical and horizontal temperature gradients.

Arbitrary orientations of the angular velocity and gravity acceleration vectors is especially important for geophysical motions when these two vectors are parallel or antiparallel only at the planetary poles. For other parts of a planet at a given latitude θ, a layer of the atmosphere or ocean may be approximated by a plane layer heated from below with the vector $\overline{g}$ perpendicular to boundaries of the layer and rotating with an angular rate $\overline{\Omega}$ over the angle $(90° - \theta)$ to $\overline{g}$, see Fig.41. Hathaway et al

(1979) showed that the most dangerous perturbations are the rolls with axes parallel to the meridian. The fastest growth of these perturbations is at the equator. The critical stability curves calculated for various latitudes θ all begin at the same value of the critical Rayleigh number and then approach the asymptotes $Ra \propto (Ta_\theta)^{\frac{2}{3}}$, where $Ta_\theta = Ta \sin^2\theta$, i.e. the angular rotation vector component, normal to the layer boundary, plays a role (the corresponding component of the Coriolis force).

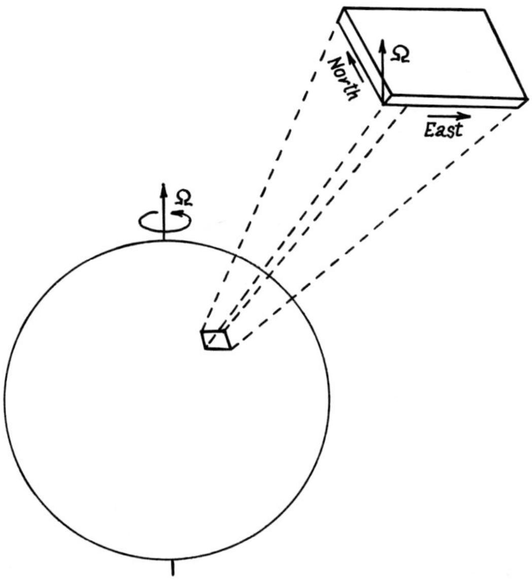

Figure 41: The geometry used in our model. A plane parallel layer of fluid is positioned tangent to a sphere at some latitude. The rotation vector is then titled from the vertical with a northward component as well as vertical component. (After Hatheway & Somerville, 1983, courtesy of Cambridge University Press).

Now, we shall consider results of computations with all components of the Coriolis force taken into account by Hathaway & Somerville (1983). For a fixed Rayleigh number $Ra = 10^4$, four cases were considered: (i) $Ta = 0, \theta = 0$, (ii) $Ta = 10^4, \theta = 90°$, (iii) $Ta = 10^4, \theta = 15°$, and (iv) $Ta = 10^5, \theta = 15°$. Results of each case served as initial conditions for the following case. Critical stability curves for various orientations of perturbations for $\theta = 15^o$ are presented in Fig.42, where also points are presented in the parameter space in which computations were performed. The most stable roll perturbations were oriented from North to South, the least stable one from East to West. For $\theta = 90°$, at the pole, the orientation of the rolls is symmetrical relative to the rotation axis and the critical stability curve is presented by a single (dotted) line for any orientation of perturbations.

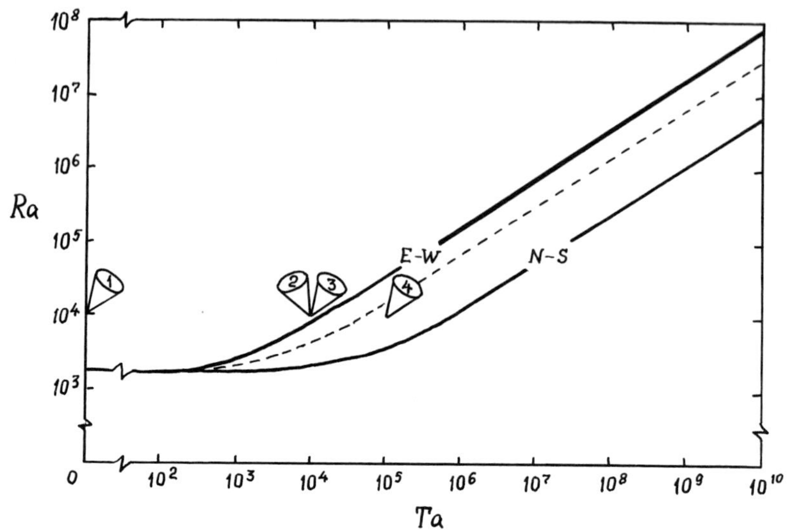

Figure 42: The stability boundaries for convective rolls at 15° latitude. The layer is unstable for Rayleigh numbers above these curves. Rolls oriented east-west (E-W) require larger Rayleigh numbers than do north-south (N-S) rolls. The dashed line represents the stability boundary for convective rolls at 90° latitude. (After Hathaway & Somerville,1983, courtesy of Cambridge University Press).

The inclined vector of the angular velocity influences all characteristics of the motions. The heat flux of the layer will be determined by a mean flow arising due to the action of the horizontal angular velocity component. An equatorward heat flux will arise which will cause a horizontal temperature gradient which in its turn will influence the mean flow. However, the periodic boundary conditions used in this work prevent the presence of a horizontal temperature gradient. The heat flux will also depend on the value of the vector $\overline{\Omega}$. As in the case of the vertical direction of $\overline{\Omega}$, here also one observes non-monotonic dependence of the Nusselt number on the Taylor number and regions on the plane $Ra - Ta$ exist where the isolines $Nu = const$ have a minimum, or from the left of the minimum the value of the Nu increases with Ta. However, in the case of the slant $\overline{\Omega}$ this increase relates to a change of the convective cell structure.

The component of the angular velocity vector along the boundaries of the layer redistributes the angular momentum in the system. The sinking fluid turns eastward, creating a downward angular momentum heat flux. This transport of the angular momentum will be partially compensated by restructuring of the pressure field for neutralization of the Coriolis force; however, the upward angular momentum flux would never appear. This effect is also absent for the motions on a sphere (Chapter 5).

Ladiede & Torrance (1991) considered the convection motions in horizontal rotating cylinder with radial and normal gravity forces. A principal motivation for this study comes from the food industry, where canned liquids are often thermally processed (sterilized or pasteurized) while rotating and moving on a conveyor belt. Two Rayleigh numbers were introduced for radial gravity and for normal gravity to choose the appropriate gravity acceleration in the definition of the Rayleigh number. The range for the main parameter was: radial and normal gravity Rayleigh numbers $\leq 10^6$, $Pr = 7$, Taylor number $Ta = \frac{4\Omega^4 r^2}{\nu^2} \leq 6.5 \cdot 10^4$, where r is the cylinder radius. The uniform volumetric heat generation was used with an isothermal outer boundary and a steady two-dimensional solution was considered. The critical condition for the radial rest state corresponds to a two-cell flow in the azimuthal plane with $Ra_{r,cr} =$ 13738. Finite-amplitude transient and steady flows are obtained with the Galerkin method for Rayleigh number ratios in the range $0.1 \leq \frac{Ra_r}{Rag} \leq 100$. This range gives the possibility to consider pure radial or normal gravity regimes as well as transition regimes. If the normal gravity dominates, then the flows are bicellular; when the radial gravity dominates, the flows tend to be multicellular. Flows driven by the radial gravity are more efficient at transporting heat than flows driven by the normal gravity.

2.13 The role of the velocity shear

The influence of the velocity shear was considered in two variants: (i) a constant shear between lower and upper boundaries was prescribed by Hathaway & Somerville (1985) and Somerville & Hathaway (1986), (ii) the shear was developed due to rotation of these boundaries with different angular velocities (see Bransvold & Vest,1974), but the difference was constant. In both cases, the problem was solved for small supercritical Rayleigh numbers, small and medium values of the Taylor number and the Reynolds (or Rossby) number. Experimental studies of the convection in the layer with one rotating boundary (upper or lower) was carried out by Berdnikov et al (1990).

The influence of shear on the Rayleigh-Benard convection in a non-rotating fluid is expressed in organizing the convective motion in the rolls oriented along the mean flow directions (see Deardorff, 1965, Ingersoll, 1966, Lipps, 1971, Hart, 1971, Clever & Busse, 1977). The motions evolve in such a way that the angular momentum is transported down the gradient and decreases the energy of the mean flow. However, the convection may in such a case take energy of the mean flow. Theory and experiment show that the critical Rayleigh number does not change under the influence of the velocity shear. For a strong heating two-dimensional rolls transform into three-dimensional cells and this transition is considered in the papers of Hart (1971), Clever & Busse (1977).

In the papers of Hathaway & Somerville (1985) and Somerville & Hathaway (1986), the shear along the layer boundaries is considered alongside with the case of different directions for $\overline{\Omega}$ and $\overline{g}$; the angle between them was 90°. At the simultaneous action of shear and rotation in order to support the constant velocity shear, it is needed to introduce into the model an external force reverse in its direction to the mean flow which would equilibrate the Coriolis force. This external force, as well as the vertical pressure gradient caused by hydrostatics and the vertical Coriolis force component, can be separated from the motion equations for the pressure in the ground state so that we can write:

$$\frac{1}{\rho_0}\frac{dP_0}{dz} = -(1-\alpha T_0)\,g + 2\Omega U_0 \cos\theta \tag{2.82}$$

where U_0 is the external velocity of the mean flow, $U_0 = Re$ at $z = \frac{1}{2}$ in non-dimensional coordinates and the Reynolds number Re is defined by the velocity shear between upper and lower boundaries ΔU as $Re = \frac{\Delta U\, h}{\nu}$. For the external force, the ground state may be written as

$$F_y = 2\Omega U_0 \sin\theta \tag{2.83}$$

and if the mean flow is given by its velocity components, i.e.:

$$U(z) = U_0(z) + \langle u(z)\rangle\,, \qquad V(z) = \langle v(z)\rangle \tag{2.84}$$

then one may determine a normalized angular momentum flux by the vector $\overline{M}$ with the following components

$$M_x = Re^{-1}\left[\frac{dU}{dz} - \langle u\,w\rangle\right], \qquad M_y = Re^{-1}\left[\frac{dV}{dz} - \langle v\,w\rangle\right] \tag{2.85}$$

$$\overline{M} = \overline{i}\,M_x + \overline{j}\,M_y \tag{2.86}$$

where $\overline{i}$ and $\overline{j}$ are the unit vectors along the coordinate axes.

In the papers by Hathaway & Somerville (1985) and Somerville & Hathaway (1986) eleven variants were considered (as continuation of the previous section 2.12 cases) at $Ra = Ta = 10^4$, $Pr = 1$: four variants for $\theta = 90°$, $Re = 50, 100, 200$ and 300 and seven variants for $\theta = 15°$, $|Re| = 50, 100$ and 200 and $Re = 300$. The module sign means that both directions of the shear were considered.

The evolution of convective motions at a change of the shear and the angle between $\overline{\Omega}$ and $\overline{g}$ is seen from Fig.43. If a small shear for $\theta = 90°$ changes the motion pattern only slightly (Fig.43a), a strong shear causes a development of roll perturbations along the shear (Fig. 43b). For a small angle $\theta = 15°$, the influence of the slant rotation overcomes the effect of the shear.

Strong changes take place with the mean flow at simultaneous action of the rotation and shear. At the absence of rotation ($Ta = 0$) a purely zonal flow arises with a shear concentrated in the boundary layers. For $Ta = 10^4$ and $\theta = 90°$, the vertical velocity profile gets closer to the superimposed external velocity profile and an additional meridional velocity component arises, but for $\theta = 15°$ the mean flow profile is almost indistinguishable from the external profile though no meridional component appears.

(a)

(b)

Figure 43: Flow field for the case of 90° latitude: (a) nonshear case with strong rotation; (b) nonrotating case with strong vertical shear (After Hathaway & Somerville,1986, National Solar Observatory, Association of Universities for Research in Astronomy, Inc.).

In a paper by Brunsvold & Vest (1973) the problem of convective instability is considered in a layer with boundaries rotating with different angular velocities. The linear stability analysis is carried out for small and medium values of the Taylor number, and for the Prandtl numbers $Pr = 0.025$, 1 and 6. An additional parameter is introduced defining inhomogeneity of the rotation of the boundaries, the Rossby number $Ro_\Omega = \frac{2(\Omega_1 - \Omega_2)}{(\Omega_1 + \Omega_2)}$ which is supposed to be small: $Ro_\Omega << 1$. For two layer values of the Prandtl number, the convection in a layer with shear has two types of instability: convective and dynamical. At small values of the Reynolds number, the stationary mode is the same as in a convective layer without shear. Rolls are stabilized and oriented in the direction of the shear. At large Reynolds numbers the shape of the critical perturbations is changing: waves of two classes, **A** and/or **B**, are caused by the mechanism of the viscous instability and depend on the Coriolis force; waves of the class **B** are related to the inflection point in the velocity profile and are almost stationary relative to the nearest boundary. A transition to another type of perturbations leads to a jump in the critical parameters of the convection onset: the wave length of the perturbation increases and the Rayleigh critical number considerably decreases.

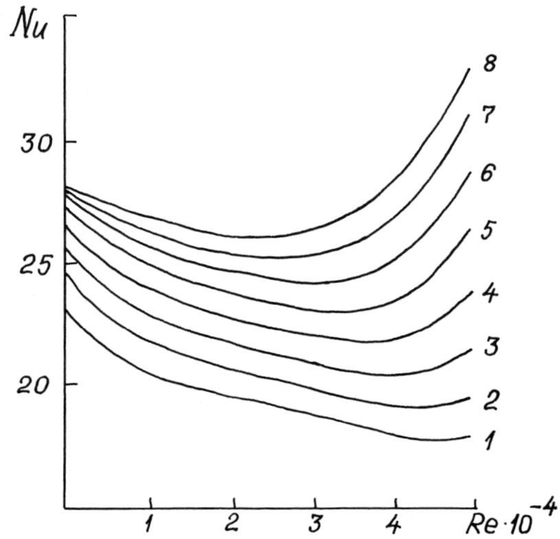

Figure 44: Dependence of Nusselt number Nu on Reynolds number Re. $Ra \cdot 10^{-7}$: 1-1.5, 2-1.7, 3-1.9, 4-2.1, 5-2.3, 6-2.5, 7-2.7, 8-2.9. (After Berdnicov et al, 1990).

At the small Prandtl number $Pr = 0.025$ and at small values of the Reynolds number, the oscillatory mode is always preferable at $Ta \geq 8000$, and at an increase of the velocity shear or of the Reynolds number, the Rayleigh number first increases and then falls down to zero where the preferable perturbation modes are waves of class **A**.

In a paper by Berdnikov et al (1990), the evolution of convective motions is studied experimentally for a plane layer with one rotating boundary. Systematic studies were carried out for the Rayleigh numbers 10^7 to $8 \cdot 10^7$ and Prandtl number $Pr = 14$ (ethanol) at the rotation of the layer upper boundary: $2.7 \cdot 10^3 \leq Re_1 \leq 1.44 \cdot 10^5$ and at the rotating lower boundary $Re_1 = 10^5$, where $Re_1 = \frac{r_0 \Omega}{\nu}$, and $r_0 = 32\,cm$ for the installation. It was obtained that the critical Reynolds number when the shear starts influencing the convective motions is $Re_{1cr} = 2 \cdot 10^3$, which agrees with an estimate of the critical number and the linear velocity of the upper boundary movement ($Re_1 = 1.25 \cdot 10^3$) and also with the condition of the same order of magnitude for the forces of inertia, viscosity and buoyancy ($Ra = Re^2\, Pr$): $Re_1 = 1.7 \cdot 10^3$ at $Ra = 5 \cdot 10^7$.

The basic flow formed in these experiments was an axisymmetric round current similar to the one forming during spin-up and considered in section 2.6. At small Rayleigh and Reynolds numbers, the flow consists of axisymmetric rolls of the same radius. At larger Rayleigh numbers, irregular motions are observed. Measurements of the heat flux through the layer revealed that in these experiments the Nusselt number first increased with the Reynolds number and then decreased. The dependence of the Nusselt number on the Reynolds number is presented in Fig.44; the existence of the minimum for the Nusselt number is evident and the position of the minimum shifts toward smaller Reynolds numbers with the Rayleigh number increasing. The decrease of the Nusselt number with the Reynolds number increase is explained by a damping of the free convection structures which corresponds to damping of the velocity normal component. At the values of the Reynolds number greater than a certain treshold value, irregulation of the flow is taking place of the forced convection nature and the heat transfer is also intensified. Appearance of a marked temperature radial inhomogeneity is noted outside the boundary layers, which takes place for $Re_1 > 6 \cdot 10^4$. This radial inhomogeneity reaches the value of 35 per cent.

Chapter 3
HORIZONTALLY TEMPERATURE-INHOMOGENEOUS ROTATING ANNULI

Convection in rotating inhomogeneously heated annuli is the second simplest case for the mutual positions of the three main determining vectors: the vector of the gravity acceleration, $\overline{g}$, and the vector of the angular rotation rate, $\overline{\Omega}$, are collinear to each other, while the vector of the external temperature gradient is perpendicular to the two. The horizontal temperature gradient in a non-rotating fluids corresponds to the convection in the layer between two vertical walls at different temperatures; for the rotating fluid, an analog to the vertical layer is the cylindrical layer.

Qualitative studies of the effects in inhomogeneously heated rotating fluids has been performed for more than two centuries and were related with attempts of explaining the circulation in the Earth's atmosphere where the meridional temperature gradient and the Coriolis force are the determining ones. A detailed review of such qualitative models of the atmospheric general circulation was presented by Lorenz (1969).

The first attempts of laboratory modeling of the influence of rotation on convection at horizontally inhomogeneous heating were carried out a long time ago by Vettin (1857) and Thomson (1892). However, the start of the regular experimental studies of such a convection in cylindrical vessels should be related to Hide who intended to model the motions in the Earth's liquid core. These studies were continued in numerous works by Hide himself and his pupils and also by Fultz with co-authors. A review of the initial period of laboratory experiments can be found in papers by Fultz (1951,1961), Hide (1958), Hide & Mason (1975). Here, we consider some basic results of the experiments on the convective motions in rotating inhomogeneously heated cylindrical vessels or annuli.

3.1 Classification of convective regimes in an annulus

The principal scheme of the experiment on modelling convection in annuli as it was originally proposed by Hide (1953) and has been used in a large number of works is shown in Fig.45. Two concentric cylinders with external radius b and internal radius a are mounted at a rotating table. The axis of symmetry of the cylinders and the axis of the whole system rotation coincide. In a space (in the form of an annulus between the cylinders) there is a working fluid. The temperature of the walls is, respectively, T_b and T_a. If for the plane rotating horizontal layer (Chapter 2) the system rotation axis could not coincide with the vessel symmetry axis (except in cases of spin-up with subsequent formation of concentric ring structures), here for the annuli with the horizontal temperature gradient the coincidence of the two axes and the same direction for both $\overline{\Omega}$ and $\overline{g}$ is principally important for studies of regular and oscillatory regimes. Further on, we shall consider in detail the condition for existence of various regimes in dependence on external parameters. Here, we only note that for regular regimes the fluid should be sufficiently deep, the radius of the internal cylinder a not too small and the rotation rate not too large.

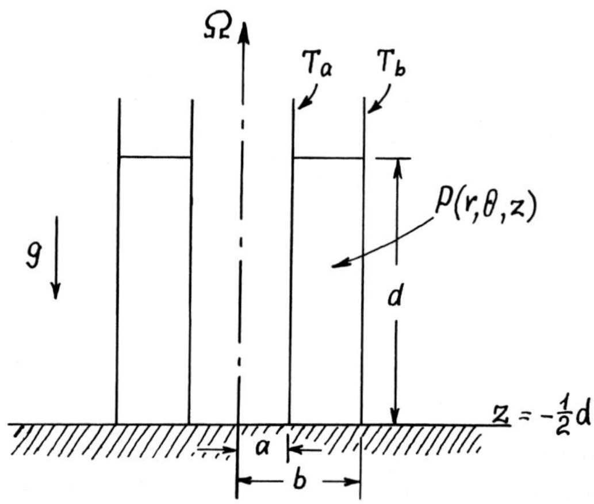

Figure 45: Schematic diagram of a rotating fluid annulus subject to a horizontal temperature gradient, drawn for the case when upper and lower bounding surfaces are horizontal. P is a general point with polar coordinates (r, θ, z) in a frame of reference rotating with the apparatus: $\overline{\Omega} = (0, 0, \Omega)$ is the angular velocity of the basic rotation; $\overline{g} = (0, 0, g)$ is the acceleration of gravity; the region occupied by the fluid is $a \leq r \leq b$, $-d/2 \leq z \leq d/2$; T_a and T_b denote the respective temperatures of the cylindrical boundaries, $r = a$ and $r = b$.(After Hide & Mason, 1975, courtesy of Adv.Phys.).

Hide (1958) proposed a classification of the convective regimes for the homogeneously heated annulus The regimes are divided into the axisymmetric (Fig.46a) and the non-axisymmetric one (Fig.46 b-d). The axisymmetric regime consists of upper and lower regimes. Non-axisymmetric regimes can be regular (46b,c) and irregular (Fig.46d). The regular regimes are stationary waves and vacillation regimes which in their turn consist of amplitude, wave and spatial vacillations. The term "vacillation" originally meant the presence of a wave of a specific form near transition to irregular regimes when regular changes in time of various characteristics of convective motions are taking place in their amplitude, wave number, or wave shape.

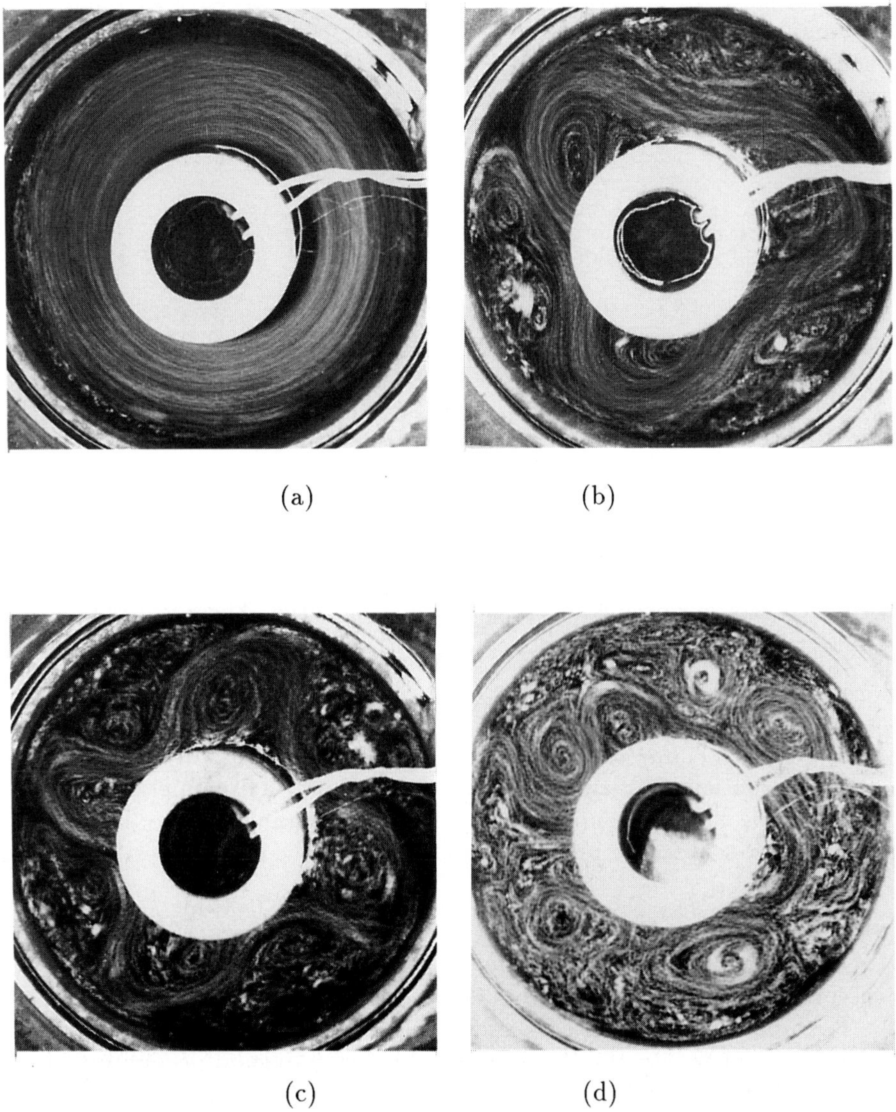

(a) (b) (c) (d)

Figure 46: Examples of the different regimes of convective motion in a rotating annulus with a horizontal gradient of the temperature: (a) axisymmetric, (b) regular waves, (c) one stage of vacillation, (d) non-regular regimes.

As for the case of a plane rotating layer considered in Chapter 2, the main external parameters determining the motion in a rotating inhomogeneously heated annulus are the Taylor number, the Prandtl number and the buoyancy flux constructed with the account of the geometry of the problem and the presence in the vessel of a stationary state which normally corresponds to the thermal wind equation. The Prandtl number has its normal form (1.22). In constructing the Taylor number, it is assumed that the influence of fluid viscosity will take place on the vessel's side walls and, therefore, more often two expressions are used:

$$Ta^* = \frac{4\Omega^2 (b-a)^5}{\nu^2 h} = Ta\,\delta, \qquad \delta = \frac{b-a}{h}$$

$$\text{or} \qquad Ta = \frac{4\Omega^2 (b-a)^4}{\nu^2} \tag{3.1}$$

where δ is the aspect ratio for the annulus.

As the non-dimensional buoyancy number, the thermal Rossby number Ro_T is usually chosen (see Eq.(1.36)), in which the ratio of the horizontal to vertical scales is taken in such a way that

$$Ro_T = \frac{gh\,\Delta\rho}{\Omega^2 (b-a)^2\,\rho} \tag{3.2}$$

A schematic diagram (after Hide) of the existence of various regimes for the convection in water with $Pr = 7$ is presented in Fig.47. A characteristic anvil-like curve separates the region of axisymmetric motions outside the curve from non-axisymmetric motions occupying the are inside the curve. There is a critical value of the Taylor number Ta^*_{cr}, such as for $Ta^* < Ta^*_{cr}$ the motion is always axisymmetric (Hide, 1969):

$$Ta^*_{cr} = (1.85 \pm 0.08) \cdot 10^5 \tag{3.3}$$

The value of $Ro_{T\,cr}$ corresponding to this value Ta^*_{cr} divides the region of axisymmetric motions into two parts: upper and lower axisymmetric regimes. The boundary of the lower axisymmetrical regime in the range of the Taylor numbers $Ta^*_{cr} < Ta^* < 2 \cdot 10^7$ can be presented in the logarithmic coordinates by the line

$$\log\left(Ro_T\,Pr\right) = (5.05 \pm 0.30) - (0.864 \pm 0.043)\,\log Ta \tag{3.4}$$

and for larger values of Ta^* this boundary becomes parallel to the axis $Ro_T = 0$. In this case, a transition takes place from the axisymmetrical regime to an irregular one.

The upper axisymmetrical regime exists always at $Ro_T > Ro_{TR}$ because at large Taylor numbers the stability curve has also a horizontal asymptote as for the case of the lower axisymmetrical regime. The existence of the axisymmetrical regime depends on its stability with respect to small perturbations and is the first question studied while considering the baroclinic instability in which the instability is related to the presence of the vertical shear of the azimuthal velocity component (the thermal wind) and with the counteracting vertical temperature gradient generated and supported by the ascending flow of the warm fluid at the external cylinder causing the meridional circulation.

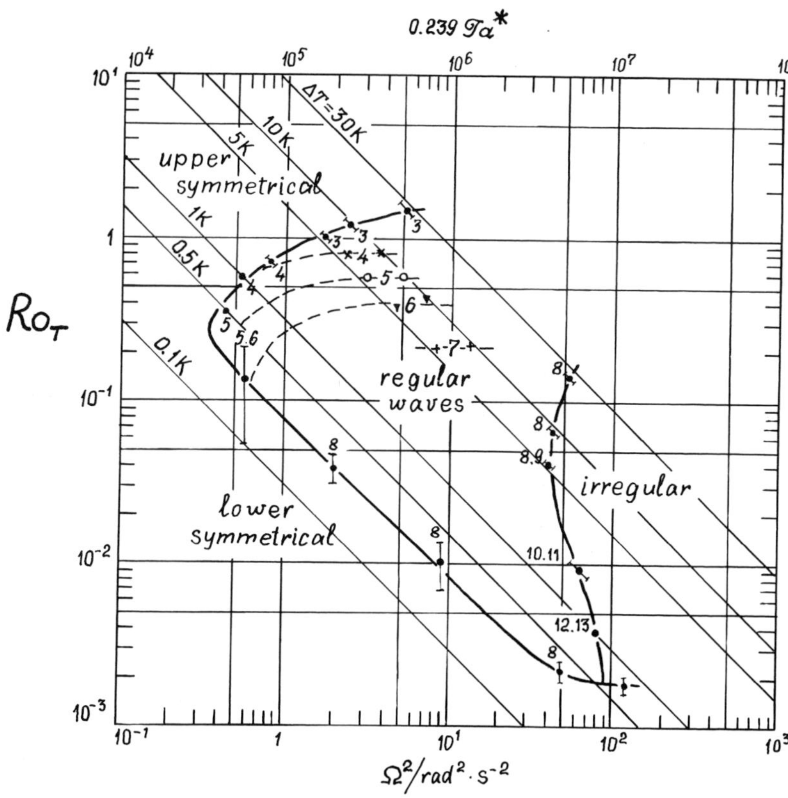

Figure 47: Typical regime diagram illustrating the dependence of the mode of convection on the two principal dimensionless parameters Ra_T and Ta^*. Working fluid is water, $a = 3.48\,cm$, $b = 6.02\,cm$, $h = 10\,cm$. (After Hide & Mason, 1975, courtesy of Adv. Phys.).

The lower axisymmetrical regime observed at small values of the thermal Rossby number Ro_T exists due to, besides the two mechanisms just described for the upper regime, transfer processes by viscosity and heat condition which damp the baroclinic instability.

Inside the anvil-like stability curve of Fig.47 the axisymmetric flow is unstable, the baroclinic instability leads to the formation of regular waves, vacillation and irregular vortex regimes. The type of the developed flow and its characteristics depend not only on the values of Ro_T and Ta^* but also on other determining parameters and in the first place on the geometry of the vessel.

The difference in the development of convective instability in the fluid at rest or in a slowly rotating fluid from the fast rotating one in which there is a geostrophic balance is illustrated in Fig.48. The lighter warmer fluid ascends along the wall of the warm cylinder, and comes to the colder cylinder wall along which it descends. In a slow rotating fluid, the angular momentum of a fluid particle proportional to $[\overline{g} \times \nabla T]$ causes the formation of axisymmetrical meridional circulation (see Fig.48a), while at large rotation rates and when there is a geostrophic balance, the gyroscopic moment proportional to $\left(2\overline{\Omega}\nabla\right)\overline{V}$ leads to the formation of a motion in the plane perpendicular to the figure's plane. Fluid particles move along trajectories slightly inclined to the horizontal at an angle substantially different from zero, but less than the slope angle for isopycnic surfaces (Fig.48b). Therefore, such a convection is often called as sloping, or slantwise convection. A particular case of such a kind of instability is a symmetrical motion in the meridional plane or baroclinic waves.

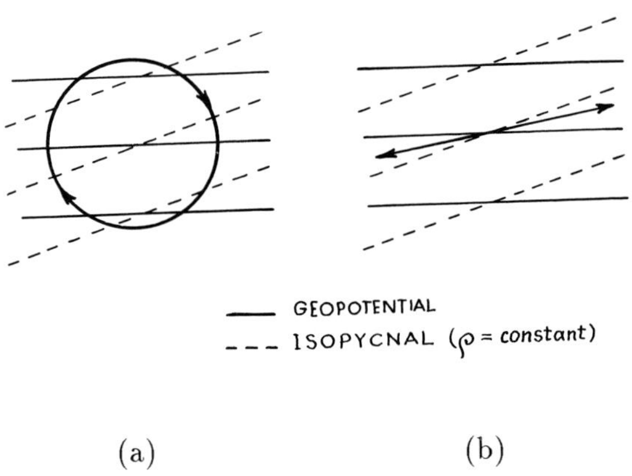

Figure 48: Illustrating sloping convection. (a) Direct meridional overturning (which occurs when Ω is sufficiently small or zero); (b) "sloping convection" in a baroclinic wave. (After Hide & Mason, 1975, courtesy of Adv. Phys.).

3.2 The axisymmetrical regime and its stability

The simplest of the regimes, the axisymmetrical one, has been studied in much details theoretically and experimentally. In the absence of rotation between the cylinders, a circulation cell is formed in which a fluid particle ascends along the surface of the heated cylinder and then descends along the surface of the cooled cylinder. At the rotation, on the fluid particle moving in the horizontal plane the Coriolis force will act and the particle will move along a spiral from the warm cylinder to the cold one while the direction of the spiraling near the upper surface is reversing at a lower part of the fluid.

For the theoretical consideration of the axisymmetrical regime an important characteristic is the Peclet number (1.21): $Pe = \frac{UL}{k}$, whose analog for the motion in an annulus is the parameter defined by Hide (1967):

$$\Pi = \frac{\alpha g|\triangle T|\nu^{\frac{1}{2}}}{8k\Omega^{\frac{3}{2}}} \tag{3.5}$$

Let us consider axisymmetric motion in an annulus. Introduce the coordinate system (x, y, z) connected with cylindrical coordinates (r, θ, z) as in Fig.45:

$$x = r - \frac{(b+a)}{2}, \quad y = \frac{\theta\ (b+a)}{2}, \quad z = z \tag{3.6}$$

Assume that $2\ (b-a) << (b+a)$ and then we may consider the system (3.6) as the rectangular one in which the equations

$$2\Omega v = \frac{1}{\rho_0}\frac{\partial P}{\partial x} - \left[\nu\,\nabla^2 u\right] + \left\{u\frac{\partial u}{\partial x} + w\frac{\partial u}{\partial z}\right\} \tag{3.7}$$

$$2\Omega u = \left[\nu\,\nabla^2 v\right] - \left\{u\frac{\partial v}{\partial x} + w\frac{\partial v}{\partial z}\right\} \tag{3.8}$$

$$\alpha g\,(T - T_0) = \frac{1}{\rho_0}\frac{\partial P}{\partial z} - \left[\nu\,\nabla^2 w\right] + \left\{u\frac{\partial w}{\partial x} + w\frac{\partial w}{\partial z}\right\} \tag{3.9}$$

$$\frac{\partial u}{\partial x} + \frac{\partial w}{\partial z} = 0 \tag{3.10}$$

where P is the deviation of the pressure from its hydrostatic distribution. The geostrophic approximation (see section 1.3) neglects the viscous and advection terms and the system (3.7)-(3.10) becomes of the geostrophic balance for an axisymmetric case:

$$2\Omega v = \frac{1}{\rho_0}\frac{\partial P}{\partial x} \tag{3.11}$$

$$2\Omega u = 0 \tag{3.12}$$

$$\alpha g\,(T - T_0) = \frac{1}{\rho_0}\frac{\partial P}{\partial z} \tag{3.13}$$

$$\frac{\partial u}{\partial x} + \frac{\partial w}{\partial z} = 0 \tag{3.14}$$

from which one can easily obtain the thermal wind equation (and the Proudman-Taylor theorem in this particular case):

$$\frac{\partial v}{\partial z} = \frac{\alpha g}{2\Omega}\frac{\partial T}{\partial x}, \qquad u = \frac{\partial w}{\partial z} = 0 \tag{3.15}$$

The heat transfer equation is written here as

$$u\,\frac{\partial T}{\partial x} + w\,\frac{\partial T}{\partial z} = k\left(\frac{\partial^2 T}{\partial x^2} + \frac{\partial^2 T}{\partial z^2}\right) \tag{3.16}$$

i.e. as a balance between the heat advection and its conduction.

Let us choose the spatial scale L in the determination of the Peclet number (3.5) as the smallest scale $L = \min\left[d, (b-a)\right]$ and consider the two asymptotics for the Peclet number. For $Pe << 1$, the heat condition prevails over the advection and (3.16) becomes

$$\frac{\partial^2 T}{\partial x^2} + \frac{\partial^2 T}{\partial z^2} = 0 \tag{3.17}$$

while at $Pe >> 1$ the advection prevails over conduction and

$$u\,\frac{\partial T}{\partial x} + w\frac{\partial T}{\partial z} = 0 \tag{3.18}$$

In Fig.49, the flow isotherms are presented schematically for the two cases. In the figure, five characteristic regions are shown: 0 is the internal non-viscous flow, 2 and 4 the side wall boundary layers, 1 and 3 the horizontal Ekman boundary layers.

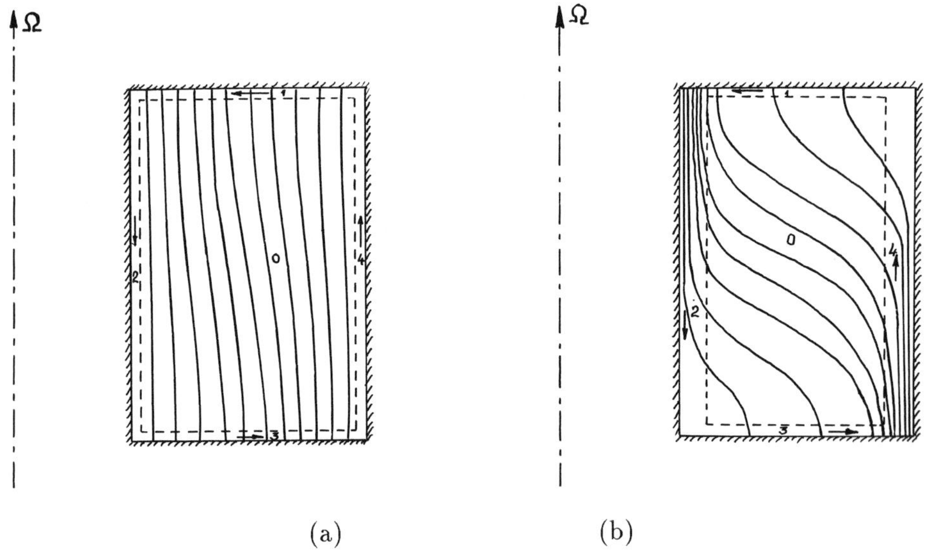

Figure 49: Illustration of isotherms: (a) low Peclet number; (b) high Peclet number. The broken lines separate schematically the cross section of the annulus into five regions: 0 - the inviscid interior; 2 and 4 - the side-wall boundary layers; 1 and 3 - the horizontal Ekman boundary layers. (After Hide, 1967, courtesy of Phys. Fluids).

The presence of the sloping isotherms for the axisymmetric regime gives a possibility to introduce (and then investigate) the relative distributions of the external temperature gradient in the radial σ_r and vertical σ_z directions:

$$\sigma_r = \frac{1}{|\Delta T|\, d}\left(\frac{\partial T}{\partial r}\right), \qquad \sigma_z = \frac{1}{|\Delta T|\, d}\left(\frac{\partial T}{\partial z}\right) \tag{3.19}$$

where $\frac{\partial T}{\partial r}$ and $\frac{\partial T}{\partial z}$ are the spatially averaged (outside the boundary layers) vertical and horizontal temperature gradients.

For the regime of heat conduction ($Pe << 1$; or $\Pi << \Pi_{cr}$, the value of which depends on the boundary conditions on the upper surface) the value of σ_r differs little from unity, as well as the value of the Nusselt number: $\sigma_r \simeq 1$, $Nu \simeq 1$. Correspondingly, $\sigma_z \simeq 0$ (see Hide,1967, Robinson,1959, Hunter,1967).

For the axisymmetrical regime with developed convection, i.e. when $Pe >> 1$, or $\Pi >> \Pi_{cr}$, Hide (1967) proposed the following expressions for σ_r, σ_z and Nu:

$$\sigma_z = 0.67 \text{ at } \Pi >> \Pi_{cr}, \sigma_r = 0.33$$

and

$$Nu = (1 + \frac{2}{3}\Pi)(\frac{1}{3} + \Pi) \text{ at } \Pi >> \Pi_{cr}$$

and also $(b-a)\sqrt{\frac{\Omega}{\nu}} = \left(\frac{1}{4}Ta\right)^{\frac{1}{4}}$. These expressions perform well at changes of Π in the limits from 15 to 42.

Investigation of the stability of the axisymmetrical motion in an annulus is the classical problem of the baroclinic instability and as such it is described in many monographs on geophysical fluid dynamics (e.g. Pedlosky, 1980). The stability of motion is considered of an inviscid fast rotating fluid in an annulus region defined as $-\frac{d}{2} \leq z \leq \frac{d}{2}$ and $a \leq r \leq b$ in the presence of constant horizontal and vertical temperature gradients $\frac{\partial T_0}{\partial z}$ and $\frac{\partial T_0}{\partial r}$. The main assumption of the theory of baroclinic instability first considered by Eady (1948) are described below. In Eq.(3.2) the density difference is decomposed into the vertical and horizontal components and instead of the non-dimensional thermal Rossby number, the new parameters are chosen: an analog of the thermal Rossby number

$$R = \frac{\alpha g d}{4\Omega^2 (b-a)} \frac{\partial T_0}{\partial r} \tag{3.20}$$

and the parameter of the static stability of the fluid:

$$B = \frac{\alpha g d^2}{4\Omega^2 (b-a)^2} \frac{\partial T_0}{\partial z} \tag{3.21}$$

The basic axisymmetric fluid motion in the coordinate system rotating with a constant angular velocity Ω at $R \to 0$ is purely azimuthal. This zonal flow changes with height according to the thermal wind equation (3.15) proportional to $\frac{\partial T_0}{\partial r}$.

The basic assumptions of the Eady model are the following:

(i) Negligence of curvature of the walls, i.e. the narrowness of the channel: $2(b-a) << (b+a)$, which easily can be reached in experiments.

(ii) Centrifugal acceleration is small compared to the gravity acceleration: $\Omega^2 r << g$. In this case one can suppose that the geopotential isolines correspond to $z = const$. The possibility to neglect the centrifugal effects was demonstrated by experiments in which the temperature difference between the cylinders was revised, i.e. the internal cylinder was heated and the outer one was cooled. The flow patterns in these two cases differ little.

(iii) Inviscid theory: $\nu = 0$. Also, the heat conduction is neglected: $k = 0$. These assumptions are fulfilled relatively well at large values of the Taylor number Ta and are applicable sufficiently well at the study of stability of the upper axisymmetrical regime. They are invalid for the lower axisymmetrical regime where the dissipative processes of viscosity and heat conduction determine the existence of the lower transition boundary and the whole anvil-like shape of the stability curve.

(iv) In the model, the external horizontal temperature gradient $\frac{\partial T_0}{\partial r}$ does not change with the fluid, i.e. the vertical shear of the azimuthal flow is constant with height. The vertical temperature gradient does not also depend on the position within the fluid and is always stabilizing: $\frac{\partial T_0}{\partial z} < 0$. The basic flow is quasigeostrophic, i.e. $Ro << 1$. These conditions are not easy to fulfill in experiments. In laboratory experiments, strong changes of the temperature gradients are observed vertically as well as horizontally. A generalization of the Eady model for the temperature gradients differing from the linear ones (see Pedlosky,1965) has been carried out only for small deviations from the linear gradients and do not help much in practice.

(v) The coefficient of the volume thermal expansion α does not depend on the temperature T. In laboratory experiments, the fluids usually used, e.g. water, have marked dependencies $\alpha = \alpha(T)$.

(vi) The slope of the isotherms for the basic state $\left(\frac{\partial T_0}{\partial r}\right)\left(\frac{\partial T_0}{\partial z}\right)^{-1}$ is of order $\frac{Rd}{b-a}$ which means that the main heat balance is due to horizontal and vertical advection. In the experiments the slope of the isotherms is always considerably larger than the value $\frac{Rd}{b-a}$. The influence of the heat conduction is increasing with $\frac{d}{b-a}$ and starting with $\frac{d}{b-a} > 2$, it gives a significant deviation between experimental and theoretical results.

(vii) Restrictions on perturbations: their amplitude should be small for applicability of the linear theory and their increment also should be small, i.e. the growth of the perturbations should be sufficiently slow in order that the geostrophic balance would be adapting to the changing flow conditions. These conditions are fulfilled well for large Taylor numbers at considering the baroclinic instability, but not so well at a comparison of the theory and experiments in the region of developed wave regimes.

Let us denote the perturbations of velocity by u_1, v_1, w_1, of pressure by p_1, of temperature by T_1, and choose according to (3.11)-(3.14) the scales $U_0 = \frac{\alpha g d}{2\Omega}\frac{\partial T_0}{\partial x}$ for

u_1, v_1 and $\frac{U_0 d}{b-a}$ for w_1, $2\Omega\rho T_0 U_0 (b-a)$ for p_1, $\frac{2\Omega U_0 (b-a)}{\alpha g d}$ for T_1, $(b-a)$ for x and y, d for z, and $\frac{b-a}{U_0}$ for t. Then the equation for the pressure perturbation can be obtained from the total system as

$$\left(\frac{\partial}{\partial t} + v_0 \frac{\partial}{\partial y}\right)\left(\frac{\partial^2}{\partial x^2} + \frac{\partial^2}{\partial y^2} + \frac{1}{B}\frac{\partial^2}{\partial z^2}\right) p_1 = 0 \tag{3.22}$$

where $v_0 = v_0(z)$ and the boundary conditions are: $w_1 = 0$ at $z = \pm\frac{1}{2}$, $u_1 = 0$ at $x = \pm\frac{1}{2}$. Represent pressure in the form

$$p_1(x, y, z, t) = Z(z) \exp\left[ik_a (y - c't)\right] \cos \pi n x \tag{3.23}$$

where k_a is the azimuthal wave number, real and positive. According to Eq. (3.22), the function $Z(z)$ satisfies the equation

$$(c - z)\left(\pi^2 n^2 + k_a^2 - \frac{1}{B}\frac{\partial^2}{\partial z^2}\right) Z(z) = 0 \tag{3.24}$$

where $c = c' - v_0(0)$ is the complex phase velocity relative to the mean flow. At the flow boundaries:

$$\left(c + \tfrac{1}{2}\right)\left(\tfrac{dZ}{dz}\right) + Z = 0 \qquad at\ z = \tfrac{1}{2}$$

$$\left(c - \tfrac{1}{2}\right)\left(\tfrac{dZ}{dz}\right) + Z = 0 \qquad at\ z = -\tfrac{1}{2} \tag{3.25}$$

The solution can be represented in the form

$$Z = K_1 \exp(\gamma_1 z) + L_1 \exp(-\gamma_1 z) \tag{3.26}$$

where $\gamma_1^2 = B\left(k_a^2 + \pi^2 n^2\right)$ which being substituted into Eq.(3.25) gives

$$c^2 = \frac{\left(1 + \frac{\gamma_1^2}{4} - \gamma_1 \coth \gamma_1\right)}{\gamma_1^2} \tag{3.27}$$

which is the dispersion relation for the complex phase velocity $c = c_R + ic_1$ and

$$\frac{K_1}{L_1} = \frac{(\gamma_1 c - 1)\cosh\frac{\gamma_1}{2} + \frac{\gamma_1}{2}\sin\frac{\gamma_1}{2}}{(\gamma_1 c + 1)\cosh\frac{\gamma_1}{2} - \frac{\gamma_1}{2}\sin\frac{\gamma_1}{2}} \tag{3.28}$$

From Eq. (3.27) and $\gamma_1^2 = B\left(k_a^2 + \pi^2 n^2\right)$ we derive a critical wave number

$$k_{a,cr} = \left[\frac{2.399}{B^2} - \pi^2 n^2\right]^{\frac{1}{2}} \tag{3.29}$$

If $k_a > k_{a,cr}$, two neutral waves are exited propagating relative to the mean flow, one in the upper part of the fluid and another one in its lower part. For $k_a < k_{a,cr}$ a combination of the two waves exists, one is a growing wave and another one is damping. The corresponding critical value of the parameter B for which the flow is stable (according to (3.29)) is

$$B_{cr} = \left(\frac{2.399}{\pi}\right)^2 \left[n^2 + \left(\frac{2\,(b-a)}{\pi\,(b+a)}\right)^2\right]^{-1} \tag{3.30}$$

which has a minimum at $n = 1$ and $2\,(b-a) << (b+a)$ is equal to

$$B_{cr\,\min} = \left(\frac{2.399}{\pi}\right)^2 = 0.581 \tag{3.31}$$

The corresponding experimental value $B_{cr\,\min} = 0.8$. The discrepancy is evidently determined by strong changes of the temperature vertical and horizontal gradients within the laboratory fluid. A deviation from linearity of the gradients in the numerical works by Williams (1974) gave a good agreement between theory and experiment.

Account for friction and/or heat exchange at the layer boundaries is a necessary condition for explaining the existence and stability of the lower axisymmetrical regime for which the viscosity and heat condition are determining processes. If one takes into account the Ekman boundary layers of the thickness $\sqrt{\frac{\nu}{\Omega}}$ at the rigid horizontal boundaries at $z = \pm\frac{1}{2}$, then only the boundary condition for the vertical velocity will change and becomes

$$w = \left(\frac{\nu}{\Omega}\right)^{\frac{1}{2}} \left(\frac{\partial v_1}{\partial x} - \frac{\partial u_1}{\partial y}\right), \tag{3.32}$$

the dispersion relation will take the form

$$c^2 + \frac{2icF_1}{\gamma_1}\coth\gamma_1 - \left(1 + \frac{\gamma_1^2}{4} - \gamma_1\coth\gamma_1 + F_1^2\right)\frac{1}{\gamma_1^2} = 0 \tag{3.33}$$

where $F_1 = \left(\frac{\nu}{\Omega}\right)^{\frac{1}{2}} \frac{d\,\gamma_1^2}{k_1 \cdot R}$ and which for $F_1 \to 0$ coincides with Eq.(3.27). Fig.50 presents a stability diagram for the axisymmetrical regime for various perturbation modes constructed by Barcilon(1964) taking into account the friction in the Ekman layers which is close in its main features to the experimental stability curve of Fig.47.

The existence of two regions of the stable axisymmetrical motions, the upper and lower one with different hydrodynamic mechanisms prevailing in the two, leads to some confusion with regard to the anvil-like stability curve. According to Fowlis & Hide(1965), the stability boundary for the upper axisymmetrical curve is universal in coordinates (Ro_T, Ta) but then the boundary for of the lower axisymmetrical regime changes with the Prandtl number and vessel geometry. Vice versa, the stability boundary of the lower axisymmetrical regime is universal in coordinates $(Pr\,Ro_T, \frac{d\,Ta}{b-a})$, but then the boundary for the upper regime shifts to the left with a decrease of the radius of the internal cylinder and an increase of the gap between the cylinders.

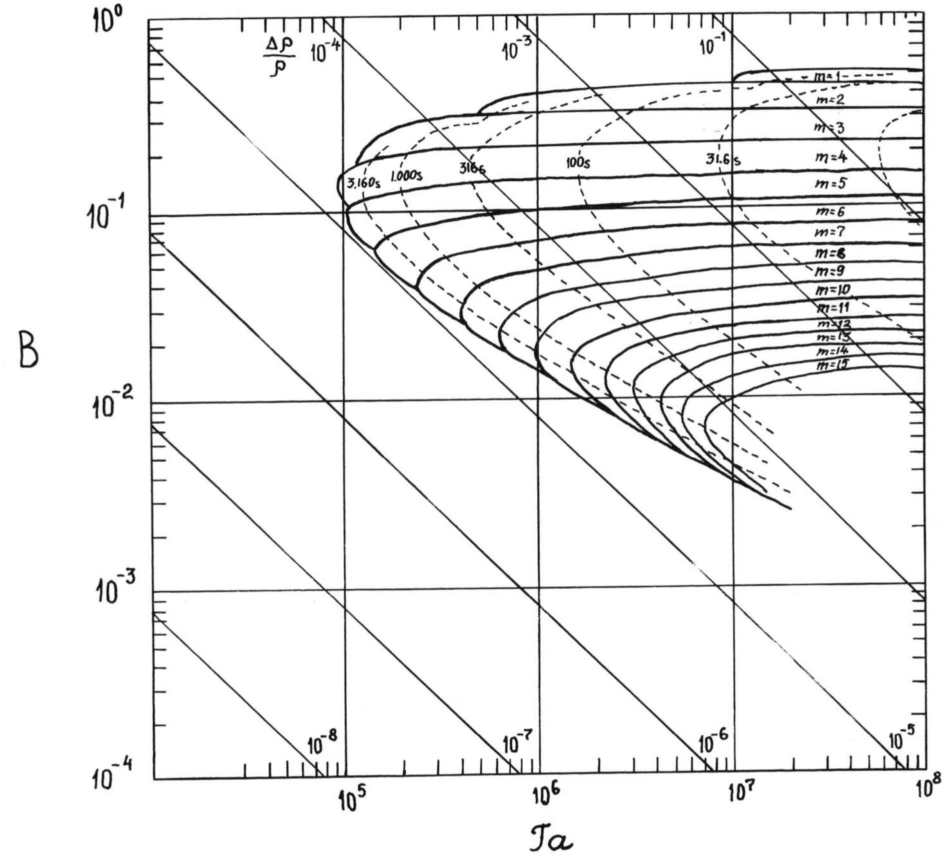

Figure 50: Regime diagram based on the theory of baroclinic instability taking into account Ekman-layer friction at the rigid upper and lower horizontal surfaces. The thick full line denotes the transition from axisymmetric to non-axisymmetric flow (marginal stability on linear theory). The region of the parameter space in which non-axisymmetric flow occurs is divided by thin full lines into separate areas each characterized by the wavenumber m which grows most rapidly; the broken lines are the corresponding contours of equal times. (After Hide & Mason, 1975, courtesy of Adv. Phys.).

3.3 Non-axisymmetrical regimes

As was said in section 3.2, the non-axisymmetrical regimes are subdivided into regular and irregular (geostrophic turbulence) areas. The regular regimes in its turn is divided into stationary waves and vacillations (auto-oscillating motions).

3.3.1 *Stationary waves*

The stationary waves are the next in complexity after axisymmetrical flows. Experimentally, in most details the dependence was studied of the azimuthal wave number m on the system's external parameters: the thermal Rossby number Ro_T, the Taylor number and on the geometry of the vessel. Besides, for a wide range of the values of Ro_T and Ta internal characteristics of the stationary waves were studied: their velocities relative to the vessel walls, the temperature and velocity fields. The wave number m was changed in experiments over a wide range and had only integer number values. The value $m = 0$ corresponds to the axisymmetric flow, $m = 1$ is the wave with one maximum (one eddy with the center not coinciding with the center of the vessel), $m = 2$ is the wave with two maxima symmetrical relative to the center of the vessel, etc. Changing the temperature difference between the annulus walls and the rotation rate, one can obtain in the experiments wave motions with the values of m in the range $m_{\text{min}} \leq m \leq m_{\text{max}}$. Minimal and maximal values of the wave number for given values of the radii a and b, which can be observed experimentally, are determined by an empirical relationship (see Hide,1970)

$$m_{\text{min}} = 0.25\,\frac{\pi\,(b+a)}{(b-a)}, \qquad m_{\text{max}} = 0.7\,\frac{\pi\,(b+a)}{(b-a)} \tag{3.34}$$

From these, one can see that at decreasing the relative value of the internal radius $\delta_0 = \frac{a}{b}$ the range of changes in the value of m is also decreasing and at sufficiently small values of δ_0 the wave motions are possible only with $m = 1$. According to Boubnov (1977,1984), however, starting with $\delta_0 = 0.2$ motions with $m = 2$ are regular but they somewhat change their pattern being non-symmetric relative the center of the vessel. For cases with small radius of the internal cylinder, the regular motions become very sensitive to the conditions at the upper boundary (free or rigid upper surface).

Maximal values of m in the experiments reached the value of 20 or more for small gaps. The existence of maximal values m_{max} for given values of a and b is due to the fact that for strictly two-dimensional flows according to the Fjortoft theorem, there is no transfer of kinetic energy by non- linear interaction to smaller scale motions without simultaneous energy transfer to larger scales. Baroclinic waves, though not strictly two-dimensional, are strongly anisotropic and regular motions can develop only when main baroclinic modes have sufficiently large scales which damps the energy transfer by non-linear interactions to other modes.

Besides by the curvature of the vessels, the existence of stationary waves is strongly influenced by the fluid depth d, which should be large enough relative to the gap width: $\delta_d = \frac{d}{b-a} > 1$. In the reverse case, the waves loose their regularity. These two geometrical constraints on the existence of experimental stationary waves are equivalent to corresponding requirements of Eady's theory (see points (iii) and (iv) of the preceding section 3.3).

For large Taylor numbers, i.e. $Ta > 2\,Ta_{cr}$ (see Eq.(3.3)), the value of m for a given Ro_T does not practically change with Ta. Due to the fact that the isotherms on the regime diagram, according to the definition of the numbers Ro_T and Ta, are inclined relative to the coordinate axis. This means that with increasing rotation rate Ω for a given temperature difference between the cylinders, the wave number m is increasing.

One of the important characteristics of the stationary waves is the existence of regions in the regime diagram in which at the same values of the external parameters Ro_T and Ta regimes exist with different wave numbers m, i.e. the intersection of regions with given wave numbers m_1 and m_2 can exist. As a consequence, a hysteresis exists, i.e. at slow changes of one external parameter not only the final value of Ro_T and/or Ta is important but also a trajectory on the plane (Ro_T, Ta) by which these values have been reached at the movement from the region where only one value, m, exists. This value can be preserved in the total area of the region of simultaneous existence of regimes with m_1 and m_2.

3.3.2 Auto-oscillating regimes - vacillations.

The next, after the stationary waves, in complexity is the motion with periodically (in time) changing wave characteristics: amplitude, phase and wave number. The diversity of the types of the auto-oscillating processes is determined by the complexity of hydrodynamic phenomena taking place at different values of the external parameters which, in the simplest case, is observed at transition from axisymmetrical motions to non-axisymmetric ones. Evidently, the complexity of wave motions at various external conditions leads to the appearance of diversified auto-oscillating regimes.

The three main types of such regimes are: amplitude vacillation, structural vacillation, and the wave dispersion (see Hide,1958,1969, Pfeffer et al,1974,1980). The amplitude vacillation is characterized by a periodic increase and decrease of a single baroclinic mode. Such types of the vacillation on the regime diagram exists near the transition from the upper axisymmetric regime and the basic motion consists of the wave with $m = m_{\min}$. The characteristic period of the amplitude vacillation is in the range 20 to 100 of the system rotation period $2\pi\,\Omega^{-1}$ and depends in large on the boundary conditions at the upper surface. For the free upper surface, the vacillation period is determined by the drift time of the wave relative to the vessel's walls at a distance of one wave length. According to experiments by Hide (1953,1969), the ratio of the amplitude vacillation period to the drift rate period was an integer. For the rigid upper surface, the drift period is much larger than the typical vacillation period but the condition that the ratio of the two periods is an integer is often conserved.

The structural vacillation is characterized by weak fluctuations of the main wave and its higher harmonics of smaller amplitude. These fluctuations are taking place mainly in phase interactions among modes and in the radial distribution of the wave energy and are of higher frequencies than the amplitude vacillation. The motion is not so perfectly periodic as in the case of the amplitude vacillation and sometimes is modulated by much longer oscillations. The main characteristic of this type of the vacillation is a periodic oscillation of the center of the wave energy (intensity) in the space relative to the mean radius of the vessel $\frac{1}{2}(b+a)$. This type of vacillation is observed between the region of regular waves and irregular regimes in the regime diagram. The main wave is developed near the maximal possible wave number $m = m_{\max}$.

The wave dispersion is characterized by the simultaneous presence of the wave number m and one or more adjacent wave numbers $m+1$, $m+2$, propagating with different phase velocities and almost constant amplitudes (see Pfeffer & Fowlis,1974).

The auto-oscillating regimes are mainly taking place at small values of $\frac{b-a}{d}$, i.e. in deep vessels and at the slopping of the isotherms near the walls in the same direction as in the main volume of the fluid. In this case the radial transfer of heat and angular momentum is done by waves which in their developed stage transfer the heat and angular momentum from the sources faster than these sources can generate them. Such a process is characteristic for oscillations in a plane horizontal fluid layer heated from below (see section 2.9), or for a plane layer with a local heat source (see Chapter 5).

3.3.3 Geostrophic turbulence

Irregular convective regimes in inhomogeneously heated annuli is investigated at sufficiently large rotation rates Ω when in the system there are conditions for geostrophic balance and, therefore, the regime is called the geostrophic turbulence. At some other values of the external parameters, irregular regimes exist when the conductions do not exist. The classification of all these regimes in the space of external parameters will be considered in Chapter 4 in which we shall discuss horizontally and vertically inhomogeneous heating in rotation fluids simultaneously.

Basic feature of the geostrophic turbulence is the existence in the system of a certain wave scale with the characteristic wave number m, determining the main part of the system energy. At this mode, the system supplies potential energy of the inhomogeneous temperature field into the kinetic energy of the flow. The dominant wave number m in the flow is equal to $m_{\max}$ (see Eq.(3.34)) and the maximum of the characteristic period in the spectrum of the temperature fluctuations is equivalent to the vacillation period and has a scale from 10 to 100 for the characteristic system rotation periods $2\pi \cdot \Omega^{-1}$, i.e. close to the vacillation period. The flow contains higher modes interacting with each other and a dominant one and there is no seemingly orderly behavior in the visual flow patterns. The kinetic energy (or temperature) frequency spectrum may be described by a power law but the exponent varies from 2.6 to 5 depending on the external parameters.

The place of geostrophic turbulence on our anvil-like regime diagram is to the right of the vacillation area.

3.4 Temperature and velocity field structure

3.4.1 Temperature characteristics

The structure of the flows in rotating annuli has been studied in much more detail and reliability than convection in the plane horizontal layer both from the experimental and theoretical point of view. As an illustration of the level of achievement in studies of the temperature field structure, one may recall the work by Pfeffer et al (1980) in which 2016 thermistors have been used to study the field at four levels within the vessel of 61 cm radius. Correlation treatment of the thermistor signals reveal sufficiently well the temperature fields and statistical characteristics of the oscillatory and irregular motions for various regimes.

For the lower axisymmetrical regime, the isotherms are practically parallel to the vertical vessel walls (see Fig.49a). For all other regimes, the isotherms are sloping (Fig.49b) which is caused by the support of the geostrophic balance, the thermal wind, within the fluid (see section 3.1). And if for the plane rotating layer at sufficiently large values of the Rayleigh number the mean non-dimensional vertical temperature gradient σ_z defined by Eq.(3.19) changes from 0 to 1 with increasing Ω for the rotating annuli at transition from the upper axisymmetric regime with increasing Ω, the values of σ_z change only weakly ($\sigma_z = 0.67$ for the upper axisymmetric regime), but the radial non-dimensional temperature gradient σ_r is considerably decreasing with increasing Ω. Such reaction of the system is connected with transportation from more complicated motion regimes to simpler ones with an increasing rotation rate Ω, which according to the Proudman-Taylor theorem makes the motion more two-dimensional in the plane perpendicular to the vector $\overline{\Omega}$ and in the annuli it leads to baroclinic instability. Such two-dimensionalization leads to small, (compared to the vertical temperature gradient) values of the horizontal temperature gradient: $\sigma_r \approx 0.33$ for the upper symmetric regimes and decreases to zero at some moments for vacillation and irregular regimes. Examples of radial profiles of azimuthally averaged temperature for the amplitude and structure vacillations and for geostrophic turbulence regime are presented in Fig.51 taken from Pfeffer et al (1980). It is worth to note that for the non-rotation case ($\Omega = 0$) for the values of the thermal forcing parameters usually used in experiments with annuli in the main volume of fluid outside the thermal boundary layers, the isotherms of the mean temperature are parallel to the horizontal surface and, therefore, $\sigma_r = 0$ (see Ketchum,1972).

Characteristic spatial distributions of the temperature within the fluid in an rotating annulus with sidewise heating are shown in Fig.52 for the upper symmetric regime at the rigid upper surface after computations by Williams(1967). These results describe well the main features of the observed motions. Characteristics features of the temperature field are the strong convergence of the isotherms in the vertical boundary layers which increases with Ω and an almost constant gradient of the mean temperature within the fluid (compare with the analogous picture for the plain convective layer). These features are common for all convective regimes and the main difference in the mean temperature characteristics among them is the increase of the mean horizontal temperature gradient σ_r (see above). For the free upper surface, due to more intense motions near it the temperature field is disorted near the free upper surface at the internal cylinder and the temperature field becomes somewhat asymmetrical.

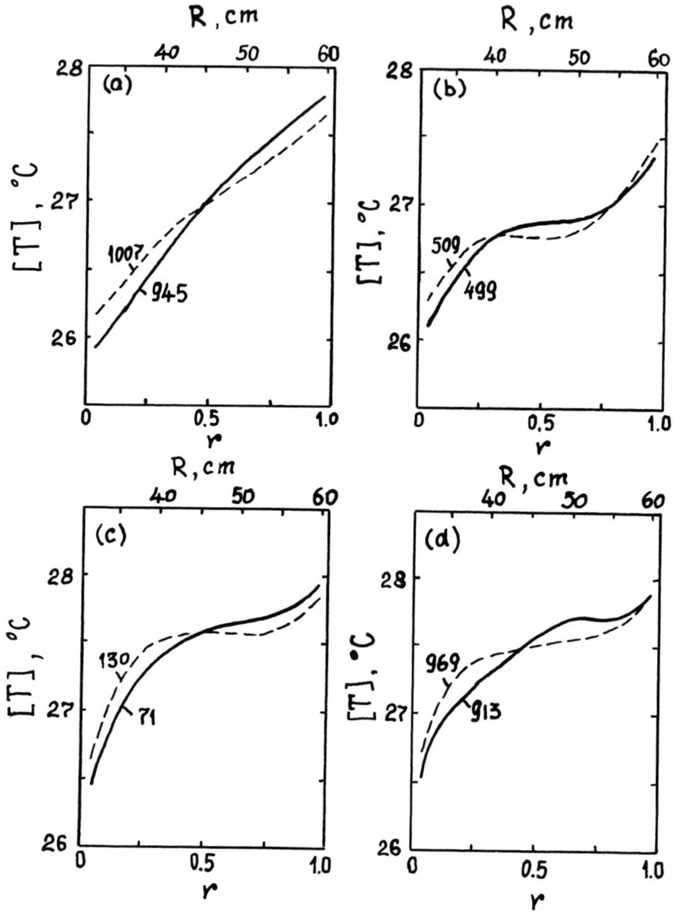

Figure 51: Radial profiles of the azimuthally-averaged temperature at mid-depth at two different times in each of the four experiments. The solid and dashed curves correspond to local (in time) maxima and minima of zonal temperature variance, respectively. The numbers refer to the number of rotations at the start of date gathering. The convective regimes: (a) 6-wave amplitude vacillation; (b) structural vacillation; (c) and (d) geostrophic turbulence. (After Pfeffer et al, 1980, courtesy of J. Atmos. Sci.).

If the averaged temperature characteristics have common regularities practically in all convective regimes, the local distributions vary considerably within a regime. E.g., according to results by Pfeffer et al (1980) the radial distribution of the mean temperature gradient $\frac{\partial T}{\partial r}$ may be a smooth almost linear dependence on radius, or could be a quadratic curve with a maximum of the temperature gradient in the center of the fluid layer for the amplitude vacillation. For the geostrophic turbulence regime, more complicated dependencies can be observed for the radial distribution of the horizontal temperature gradient, in particular a transition at the change of the rotation rate Ω from the curve with one minimum at the vessel center to a curve with central maximum and two minima near the fluid boundaries.

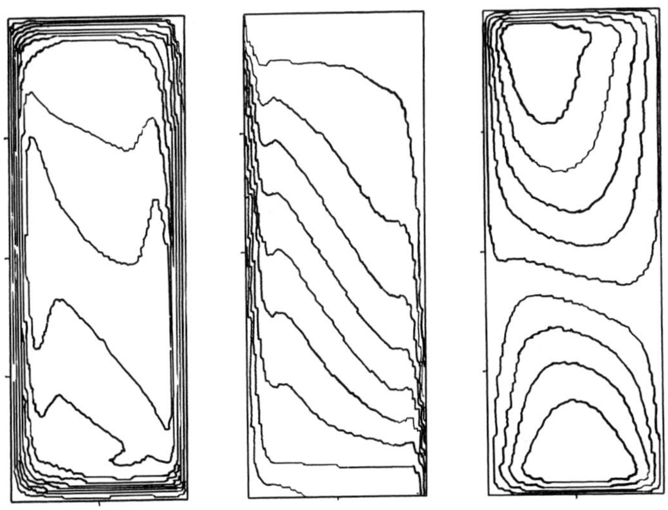

(a) (b) (c)

Figure 52: Typical steady-state contours of the stream function (a), normalized temperature (b) and zonal velocity (c) for axisymmetric flow in a rotating fluid annulus subject to a horizontal temperature gradient when the upper surface is free. (After Williams, 1967, courtesy J. Atmos. Sci.).

The distribution of the temperature fluctuation variance, σ_T^2, is determined by strong flows in the boundary layers near the cylinders. The basic distribution is a smooth curve with a maximum near the center of the gap between the cylinders. At approaching the upper boundary, the value of σ_T^2 is increasing and the curve has a sharper maximum near the gap center. For some cases of motion in the regimes of the structure vacillation and geostrophic turbulence, in the lower part of an annulus the distribution $\sigma_T^2(r)$ has two maxima near the cylinders and a minimum in the center of the gap, which reminds us for the distribution of the temperature fluctuation variance for a plane rotating fluid layer (Chapter 1). However, while approaching the mean level of the fluid layer this distribution transforms again into a curve with one maximum.

Now, we shall consider spectral characteristics of the various convective regimes. The spectrum of the temperature fluctuations reflects basic features of the regimes: presence or absence of prominent modes and their relative role. Fig.53 presents the spectral energy density S_T vs. the wave number m for two types of vacillation regimes and for two regimes of geostrophic turbulence. The main features of the amplitude vacillation (Fig.53a) is the presence of well defined maxima at the wave numbers nm, where $n = 1, 2, 3...$ and the basic number $m = 6$ and the near absence of any other modes in the system. For the structure vacillation (Fig.53b), the maxima are a bit less prominent for the basic mode $m = 5$ and for modes nm (n =2,3,...) but other modes are also present in the spectrum. At transition to the regimes of the geostrophic turbulence, the spectrum becomes smooth with one maximum (see Fig.53c,d).

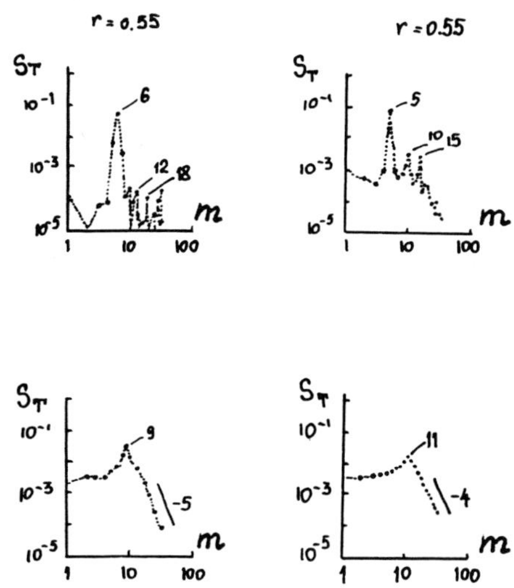

Figure 53: Time-averaged azimuthal temperature variance spectra for the experiments of Fig.51 (After Pfeffer et al, 1980, courtesy of J. Atmos. Sci.).

In Fig.53, the spectra are shown measured at the center of the gap and at a half the fluid layer depth. The figure is taken from the paper by Pfeffer et al (1980), in which the spectra are presented for three different horizontal levels and for three different points of the radius within the gap. The main features of the spectra change only weakly in dependence on the position within the main bulk of the fluid. We only note a change in the maximum position of the spectrum in Fig.53d, which near the cold cylinder corresponds to $m = 9$ but starting with the center of the gap it shifts to $m = 11$. The spatial pattern of radially averaged temperature fluctuations for certain wave numbers shows that modes with different values of m for the geostrophic turbulence regime have different spatial distributions at various time moments.

In the same work by Pfeffer et al (1980), distributions of amplitude of r.m.s. temperature fluctuations σ_T are presented for separate modes developing in the system in dependence on time which shows that for the amplitude vacillation a periodic development of one main mode and one additional mode takes place. In the structure vacillation this periodicity is breaking down though one observes a strong prominence of the basic mode and of additional ones but also there are many overtones of smaller intensities, for the irregular geostrophic turbulence the prominence of main modes is rather weak on the background of practically all modes under consideration.

The frequency spectra of the temperature fluctuations have been considered by Buzina et al (1990). For the regime of geostrophic turbulence, it was obtained that the spectral shape did not change in the limits of the measurement accuracy at changing the rotation rate and the temperature difference between the cylinders. The shape of these spectra is similar to the geostrophic convection spectra in the plane horizontal rotating layer (see Fig.33). However, due to lack of the numerical values for the spectra it is not possible to present these spectra in the universal coordinates of section 2.10 and an attempt to use the results of Fig.13 of this paper proved to be unreliable. A new estimate is also required for the most probable oscillation frequency. However, the similarity of the shapes of the spectra gives hope that the rotation does not significantly change the main spectral features for turbulent regimes in closed cavities. At this point, it is proper to recall Fig.30 where the spectra in universal non-dimensional coordinates are the same for two cases: of vertical heating (the constant temperature at a vertical plate) and horizontal heating. It is natural to suppose that an analogous similarity should be observed for the thermal turbulence for the conditions of geostrophic balance.

Buzina et al (1991) in their paper paid the main attention to finding "power laws" for the temperature frequency spectra. In the lower part of the spectra, they found the power exponent -4. But the spectral interval is rather short and the exponent is changing for different frequency intervals, as it was the case in Fig.33.

3.4.2 Velocity field characteristics

Experimental measurements of the frequency field in rotating annuli is much more complicated compared with temperature measurements. This is because the characteristic velocity values are typically of the order of a few millimeters per second and for this range there are no electronic sensors, similar to thermocouples or thermistors for the temperature. The most reliable and the most time and effort consuming method for velocity is flow visualization and the subsequent fixing the flow patterns in a fixed time interval. Any measurement of velocity inside the annuli requires installation of transparent windows in the copper or brass cylinder walls which poses severe complications for carrying out the experiments. Measurements of velocities at the surface gave only restricted information and could be performed only for the free upper surface.

The main feature of the velocity field is the presence of the zonal motion at opposite directions in the upper and lower parts of the vessel. This is a consequence of the presence of the Coriolis force acting on the convective cell with ascending motion near the heated cylinder and descending motion near the cooled cylinder with opposite flows in the radial direction in the upper and the lower parts of the vessel. A typical zonal velocity distribution for the axisymmetric regime is presented in Fig.52. Such an almost symmetric distribution for the case of a rigid upper surface becomes highly asymmetric for the case of a free upper surface when due to the absence of friction at the free surface the flow in the upper part becomes several times more intense than the counter flow in the lower part of the vessel.

The main characteristics studied in many experiments is the angular velocity of the drift of the wave pattern in the regime of the regular waves. It was usually determined as $\omega_d = \frac{2\pi}{mT_v}$, where T_v is the period of the wave with the wave number m. In the wave regime near the transition from an axisymmetric regime, the drift angular velocity is proportional to the value of the external thermal wind $\alpha\,\Delta T\,\Omega^{-1}$ (see Green, 1970):

$$\omega_d = C_\omega \left(\frac{\alpha\,\Delta T}{\Omega}\right)^n \tag{3.35}$$

The values of the coefficient C_ω and the exponent n depend on the character of the boundary condition at the upper surface. For the rigid upper boundary, $C_\omega = 13.3 \pm 6.2$, $n = 0.88 \pm 0.07$ and for the free surface $C_\omega = 2.4 \pm 1.2$, $n = 1.0 \pm 0.1$. The asymmetry of the temperature field for the free upper surface leads to an increase of the drift velocity by almost an order of magnitude.

At a certain moment inside the region of the regular waves on the regime diagram, the increase of the drift velocity is slowing down with an increase of the thermal wind velocity and a sufficient scatter of experimental points is observed.

3.5 Heat and angular momentum transfer

Studies of heat transfer in rotating annuli can be divided into two types: (i) hydrodynamical, in which the heat flux averaged over space and time, the Nusselt number, was studied in a wide range of the external parameters and (ii) "atmospheric", where for certain values of the external parameters corresponding to the general circulation attempts were made to connect this transfer with the main internal characteristics of the system.

A characteristic dependence of the Nusselt number on the system angular velocity Ω is presented in Fig.54. For the region of the axisymmetric regime, one observes an almost square root decrease of the Nusselt number for an increase of Ω. The behavior of the Nusselt number for the axisymmetric regime can be obtained (see Hide & Mason,1975) assuming that the convective transfer is taking place only in the viscous Ekman boundary layers of width $\delta_E = \left(\frac{\nu}{\Omega}\right)^{\frac{1}{2}}$; outside these layers, the geostrophic balance should take place and at the boundary of the layer the characteristic horizontal velocities can be written as

$$u \propto v \propto \frac{\alpha g \sigma_r d \, \triangle T}{4\Omega\,(b-a)} \tag{3.36}$$

and the corresponding velocity contribution for the heat transfer between the cylinders is

$$\frac{H_q}{\rho c_p} \propto u \, \delta \, \sigma_z \, \triangle T$$

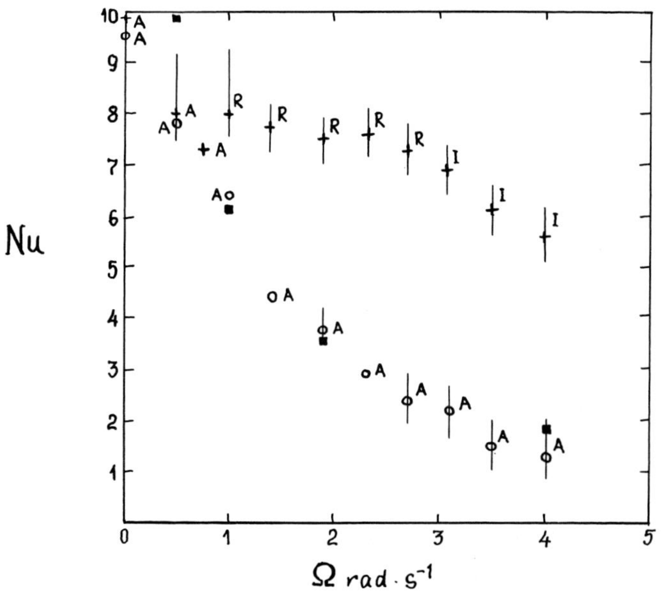

Figure 54: Dependence of the Nusselt number Nu on the rotation rate Ω. A denotes axisymmetric flow, R regular baroclinic waves and I irregular waves.(After Hide & Mason,1975, courtesy of Adv. Phys.).

Measuring the total heat transfer between the cylinders $\pi\,(b+a)\,H_q$, the nondimensional heat flux, i.e. the Nusselt number, can be written as

$$Nu = \frac{H_q}{k\,\rho\,c_p\,d\,(b-a)\,\triangle T} \tag{3.37}$$

which for the axisymmetric regime can be represented as

$$Nu \simeq \sigma_r\,(1+\sigma_z\Pi) \tag{3.38}$$

where σ_r and σ_z are non-dimensional radial and vertical temperature gradients defined by Eqs.(3.18) and (3.19) and Π is analog of the Peclet number according to Eq.(3.5). The dependence (3.38) is well confirmed by experiments (Fig.54, lower points with index A).

For the wave regime, the Nusselt number changes only weakly with Ω and in the regime of the geostrophic turbulence a strong practically linear decrease of the Nusselt number is observed with Ω increasing (see Hide & Mason,1975): $Nu \propto \Omega^{-1}$, while in the axisymmetric regime (see above): $Nu \propto \Omega^{-\frac{1}{2}}$.

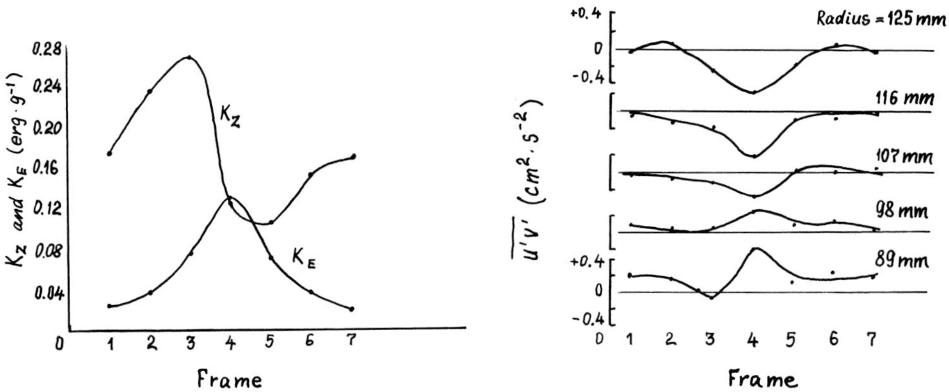

(a) (b)

Figure 55: The mean K_z and eddy kinetic energy K_E (a), eddy moment transport $\overline{u'v'}$ (b) for one amplitude vacillation cycle (After Fitzjarrald, 1982, courtesy of J. Phys.).

Spatial and time distributions of some energy characteristics for wave and irregular regimes ware considered by Pfeffer et al (1980), Fitzjarrald (1982). A determination of the energy and angular momentum transfer in the vacillation cycle was carried out by measuring the velocity field by a track method. One level was illuminated by a light knife in the fluid at a distance from the bottom of 3/4 of the total depth and, using a track length, every 10 seconds the visualization particle velocity was measured in the illuminated layer. The mean kinetic energy K_z and the eddy kinetic energy K_E were determined as

$$K_z = \frac{\int (\overline{u}^2 + \overline{v}^2)\, r\, dr}{2 \int r\, dr} \tag{3.39}$$

$$K_E = \frac{\int \left(\overline{u'^2} + \overline{v'^2}\right) r\, dr}{2 \int r\, dr} \tag{3.40}$$

where the fluctuations $u' = u - \overline{u}$, $v' = v - \overline{v}$ and the averaging is carried out at a given radius.

The values of K_z and K_E determined for one vacillation cycle are presented in Fig.55a . It shows that in the middle of the cycle a sharp decrease of the mean kinetic energy is observed, simultaneously the eddy component of the kinetic energy is minimal, but then the value of K_E starts to decrease. The corresponding angular momentum transfer by the eddy component $\overline{u'v'}$ for several values of the annulus radius is presented in Fig.55b, where one can clearly see the angular momentum transfer into the jet stream in the central part of the region between the cylinders (here $b = 15$ cm, $a = 7.5$ cm).

An attempt to find the best relationship between the heat flux transported by the motion eddy component and the mean temperature gradient within the fluid was done by Pfeffer et al (1980b). Usually, for a parametrization of this relationship the following dependence is used:

$$\overline{v'T'} = C_{vT} \left(\frac{\partial \overline{T}^p}{\partial y} \right) \tag{3.41}$$

where $p > 0$ and C_{vT} is also a positive value which either weakly depends on the rotation rate, or is inversely proportional to Ω in some positive degree (see Green, 1970, Stone, 1972).

Experimental results by Pfeffer et al (1980b) show that the observed dependence of the eddy flux heat on the mean temperature gradient is much more complicated than that represented by Eq.(3.41). Fig.56 gives dependencies of the time averaged radial heat flux $\left[\overline{v'T'}\right]_{MM}$ at the middle level and in the center of the gap between the cylinders (MM) in dependence on the internal temperature difference $\left(\overline{\Delta T}\right)_{MM}$ (i.e. internal gradient) measured at the same place for various values of the external temperature difference and the rotation rate. For a given temperature difference between the cylinders, the eddy heat flux first increases with Ω and then after a maximum starts to decrease. The system goes through the regimes of upper symmetric amplitude and structure vacillation and geostrophic turbulence. The dotted line in Fig.56 defines a linear dependence of the Eq.(3.41) type but only for maximal values of the eddy heat flux. In the paper by Pfeffer et al (1980b), a parametrization was proposed which satisfied better the experimental results also in other points:

$$\frac{\partial H'}{\partial t} = k_G\, G' + k_H\, H' + k_0 \tag{3.42}$$

where

$$H' = [v'T'] - \left[\overline{v'T'}\right], \; G' = \frac{(b-a)}{\Delta T}\left(\frac{\partial T}{\partial r} - \frac{\partial \overline{T}}{\partial r}\right),$$

and k_G, k_H, k_0 are numerical coefficients. A theoretical model constructed by Pfeffer & Barcilon (1978) includes effects of weak non-linearities. It shows that if an analysis is carried out for separate modes, then the eddy heat transfer for every mode increases with Ω at a fixed value of the temperature difference between the cylinders, but there is a sharp fall of the heat flux at transition to a mode with a higher wave number.

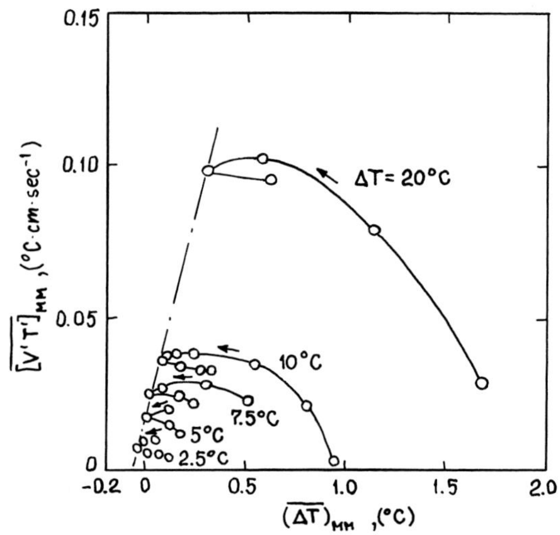

Figure 56: Graph showing the relationship between time-averaged radial eddy heat flux and internal radial temperature difference at mid-depth and mid-radius. Solid lines connect points corresponding to different experiments with the same imposed temperature contrast. Arrows indicate the direction of decreasing thermal Rossby number Ro_T along diagonals on the regime diagram at constant imposed ΔT. The thin broken line connects points with minimum $\overline{(\Delta T)}_{MM}$ at different imposed ΔT. (After Pfeffer et al, 1980b, courtesy J. Atmos. Sci.).

The radial distribution of the eddy heat flux reminds us of distributions for the temperature fluctuation variance: for the regular regimes, there is a characteristic maximum at the center of the gap between the cylinders and for the structure vacillation and geostrophic turbulence complicated dependencies are observed with two or three extrema.

3.6 Various Prandtl numbers

The regime diagram of Fig.47 is obtained for the water with the Prandtl number $Pr \approx 7$. The number water depends rather strongly on temperature, being 7 for $20\,°C$, and decreases to 3.5 for $90\,°C$. The other fluids have a wide spectrum of values. In sections 3.1-3.5, the results were presented which being obtained for different values of the Prandtl number are general for a wide range of external parameters. Now we consider after Fein & Pfeffer (1976) the practical changes which take place at changes of the Prandtl number for sufficiently wide limits from 0.025 to 63. Recall that *Pr,* as the ratio of kinetic viscosity to the molecular diffusivity, is characterizing the relative role of these two molecular processes in a field.

The regime diagram for three different fluids, i.e. mercury with $Pr = 0.0246$, water, $Pr = 7.16$ and silicon oil, $Pr = 63$ in the three-dimensional space of parameters (Ro_T,Ta,Pr), is presented in Fig.57. Main features of the diagrams consist in significant changes of regime boundaries as well as in disappearance of some and appearance of new regimes.

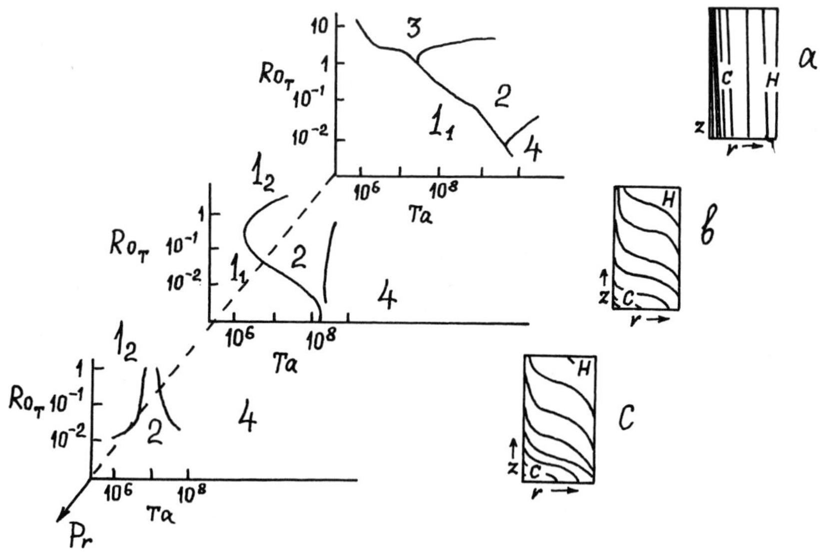

Figure 57: The regime diagrams for silicone oil, water and mercury plotted in the three-dimensional space (Ro_T, Ta, Pr). To the right of the regime diagrams show meridional cross-sections of the zonally averaged temperature measured at $\Delta T = 5\,°C$ approximately mid-way across the regular wave regime for each of the three fluids. Regimes: 1_1- lower symmetric, 1_2- upper symmetric; 2 - regular waves; 3 - thermal turbulence (non-geostrophic); 4 - irregular waves (geostrophic turbulence). (After Fein & Pfeffer, 1976, cortesy of Cambridge University Press).

The position of the boundary between regular waves and irregular geostrophic regimes noticeably change. The critical value of the thermal Rossby number Ro_T at small Pr first increases with Taylor number, at intermediate values of Pr it does not change appreciably with Ta and, finally, for large Pr it decreases with Ta increasing.

3.6.1 Small Prandtl numbers: Pr<< 1

For small values of Pr (mercury) the regions of the regular waves and geostrophic turbulence exist at sufficiently large values of the Taylor number Ta and thermal Rossby number Ro_T. Consequently, the area of the existence of the lower axisymmetric regime increases considerably due to high thermodiffusivity of the fluid. The main feature of the temperature field is that isotherms are parallel to the side walls of the cylinders. Such a distribution of isotherms is more or less conserved for other regimes also and the maximal deviations of the isotherms from the vertical does not exceed 20°.

For small Prandtl numbers, there is no upper axisymmetric regime. Instead of it, an irregular regime is observed called by Fein & Pfeffer (1976) "ageostrophic turbulence" because the thermal Rossby number is of order or larger than unity (up to 10) and the Richardson numbers are small ($Ri \approx 1$). The transition from the lower axisymmetric regime to ageostrophic turbulence takes place abruptly at a change of external parameters, while the hysteresis phenomenon is not observed at this boundary. The case is different from the boundary with regular waves which is strongly spread in the external parameter space. In the region of geostrophic turbulence, the temperature variance σ_T^2 does practically not depend on the temperature difference between the cylinders and depends linearly on the angular rotation rate Ω. Here we may recall that the dependence $\sigma_T^2 \propto \Omega$ was observed for the irregular geostrophic convection in the plane rotating fluid layer. However, there according to Eq.(2.67) also a proportionality to the heat flux occured coming through the fluid layer.

At the transition from ageostrophic turbulence to the wave regime and also in the region of the purely wave regime, an interesting phenomenon appears: a change of the direction for the wave drift. For water (section 3.4), the waves at cold central and warm external cylinders drifted relative the vessel walls in the direction of the general rotation of the system. For mercury, at a given temperature difference $\triangle T$ and at an increase of Ω, first waves arise superimposed on turbulence and moving in the direction opposite to the general rotation, then the waves become stationary relative to the walls and with a further increase of Ω the waves start to drift in the direction of the rotation.

3.6.2 Large Prandtl numbers: Pr>> 1

The notable feature of the regime diagram for large Prandtl numbers ($Pr = 63$ for silicon oil) is the absence of the lower axisymmetric regime and there is no anvil-like stability curve for the axisymmetric regime. The transition from the upper axisymmetric regime to the wave regime at large values of the thermal Rossby number Ro_T

depends only weakly on the Taylor number Ta. For $Ro_T << 1$, instead of the lower axisymmetric regime, the wave regime is formed in which, instead of a subsequent increase of the wave number m at decreasing Ro_T (for a fixed Ω) as for water, closed regions of parameters appear with maximal values of the wave number m. A qualitative description of the transition boundary from axisymmetric to the wave regime for a changing the Prandtl number value can be obtained using a model which includes accounting for the slope of the geopotential isolines which arises in the rotating system due to the action of centrifugal forces.

The transition boundary from the wave motions to the geostrophic turbulence can be presented by a curve in which the slope has changed in comparison with the case of $Pr << 1$ and with the Taylor number increasing, the values of the critical thermal Rossby number Ro_T are decreasing for this transition. Besides, the region of the geostrophic turbulence itself shifts towards smaller values of the Taylor number.

The thermal structure of the flows at large values of the Prandtl number is similar to the structure in water. Strongly sloping isotherms are observed, the temperature oscillation amplitude in a wave for a fixed temperature difference between the cylinders is linearly increasing with the angular rotation rate Ω.

3.7 Some non-linear models

One of the main aims of the analysis of the convective motions in rotating annuli is the study of the mechanism of the baroclinic instability, one of the key mechanisms of the large scale atmospheric motions. Therefore, in the main theoretical investigations of the atmospheric circulation a comparison was made with results of studies on convection in the annuli with the prescribed temperature difference between the cylinders.

The development of theoretical models was in two main directions:

(i) Development of Eady's model (section 3.3) accounting for non-linear interaction between various modes with different values of the wave number m, often with consideration of dissipation in the Ekman layers. In this direction there is a possibility to elucidate the role of the vertical structure in the process of the interactions; however, this brings about large mathematical difficulties.

(ii) Development of two-layer models which mathematically are much simpler but allow one to consider much more complicated interactions at the prescribed vertical structure. Here, we shall consider some results of these two directions in application to convection in annuli. More extended reviews can be found in papers by Hide & Mason (1970), Pedlosky (1977), Barcilon & Drasin (1984).

Two-layer geostrophic models obtained their logical completion in a cycle of papers by Lorentz (1960,1962,1963,1984). He was able to account for the non-linear interactions of various modes and to describe, at least qualitatively, practically all complicated types of motion and their changes observed in laboratory experiments. However, this too simplified description of the heat and momentum transfer between the layers and at the boundaries made quantitative comparison between the model and experiments impossible. These two-layer models were obtained as the simplest case of a multilayer models proposed by Lorentz (1960) for consideration of energy exchange in the atmosphere. The motion equation system (1.5)-(1.7) is added by the equation of state and the hydrostatic equation. The final model includes three equations for vorticity, continuity and heat balance.

For the two layers, stream functions for the non-divergent solenoidal part of the horizontal motion are introduced: $\psi+\tau$ and $\psi-\tau$ for the upper and lower layers, $\theta+\sigma$ and $\theta-\sigma$ for the temperature in the layers, $-\chi$ and χ for the velocity potentials for the potential part of the horizontal motion. At the constant Coriolis parameter f_c, the model equations can be written as

$$\frac{\partial \nabla^2\psi}{\partial t} = -J(\psi, \nabla^2\psi) - J\left(\tau, \nabla^2\tau\right) \tag{3.43}$$

$$\frac{\partial \nabla^2\tau}{\partial t} = -J(\psi, \nabla^2\tau) - J(\tau, \nabla^2\psi) + f_c\,\chi \tag{3.44}$$

$$\frac{\partial \theta}{\partial t} = -J(\psi, \theta) - J\,(\tau, \sigma) + \nabla(\sigma\nabla\chi) \tag{3.45}$$

$$\frac{\partial \sigma}{\partial t} = -J(\psi, \sigma) - J\,(\tau, \theta) + \nabla\theta\,\nabla\chi) \tag{3.46}$$

where t is time and $J(a,b) = \frac{\partial a}{\partial x}\frac{\partial b}{\partial y} - \frac{\partial b}{\partial x}\frac{\partial a}{\partial y}$ is the Jacobian. The equation system is closed by the thermal wind equation, the relationship between the temperature gradient and wind shear

$$\nabla^2\theta = B_T\,\nabla^2\tau \tag{3.47}$$

where $B_T = \frac{\alpha g}{f_c}$ (α is the thermal expansion coefficient). To obtain a simplified system, the Galerkin expansion is used on the system of orthogonal functions $F_0, F_1, F_2, ...$, which satisfy the following conditions:

$$L^2 \cdot \nabla^2 F_i = a_i\,F_i, \qquad i = 0, 1, 2... \tag{3.48}$$

where L is a constant of the dimension of length, and

$$\frac{\partial F_i}{\partial S} = 0, \qquad i = 0, 1, 2... \tag{3.49}$$

is the tangential derivative everywhere at the boundary and

$$F_i F_j = \delta_{ij} \tag{3.50}$$

where δ_{ij} is the Kronecker symbol: $\delta_{ij} = 0$ at $i \neq j$ and $\delta_{ij} = 1$ at $i = j$. We assume $F_0 = 1$, i.e. $a_0 = 0$. The Jacobian from the two orthogonal functions satisfies the relationship

$$L^2\,J(F_j, F_k) = \sum_{i=0}^{\infty} c_{ijk}\,F_i \tag{3.51}$$

where

$$c_{ijk} = L^2\,\overline{F_i\,J(F_j, F_k)} \tag{3.52}$$

where the observed line means averaging. Expansions are introduced, i.e.

$$\psi = L^2 f_c \sum_{i=1}^{\infty} \psi_i F_i, \qquad \theta = B_T L^2 f_c \sum_{i=1}^{\infty} \theta_i F_i,$$

$$\tau = L^2 f_c \sum_{i=1}^{\infty} \tau_i F_i, \qquad \sigma = B_T L^2 f_c \sum_{i=1}^{\infty} \sigma_i F_i, \qquad \nabla^2 \chi = f_c \sum_{i=1}^{\infty} \omega_i' F_i \tag{3.53}$$

Substituting expansions (3.53) into the system (3.43)-(3.46) and assuming that the temperature perturbation depends only on time, i.e. $\sigma = \sigma(t)$, we obtain the spectral system of equations:

$$\frac{\partial \psi_i}{\partial t} = \frac{1}{2} \sum_{j,k=1}^{\infty} a_i^{-2} \left(a_j^{-2} - a_k^{-2}\right) c_{ijk} \left(\psi_j \psi_k + \tau_j \tau_k\right) \tag{3.54}$$

$$\frac{\partial \tau_i}{\partial t} = \frac{1}{2} \sum_{j,k=1}^{\infty} a_i^{-2} \left(a_j^{-2} - a_k^{-2}\right) c_{ijk} \left(\tau_j \psi_k + \psi_j \tau_k\right) \tag{3.55}$$

$$\frac{\partial \theta_i}{\partial t} = \frac{1}{2} \sum_{j,k=1}^{\infty} \left(\theta_j \psi_k - \psi_j \theta_k\right) + \sigma_0 \omega_i' \tag{3.56}$$

$$\frac{\partial \sigma_0}{\partial t} = \frac{1}{2} \sum_{j,k=1}^{\infty} \theta_i \omega_i' \tag{3.57}$$

$$\theta_i = \tau_i \quad \text{if } a_i \neq 0. \tag{3.58}$$

The basic problem of the studies is now the determination of the expansion coefficients a_i and c_{ijk} and the solution of the ordinary differential equation system (3.54)-(3.58). The system does not have dissipation forces. There are several ways of their introduction. It is usually assumed that the friction force of the lower boundary is proportional to the velocity in the lower layer and the force between the layers is proportional to the velocity shear on the boundary between the layers. Similar assumptionsexist for the heat exchange at the lower boundary: it is proportional to the temperature difference between the fixed temperature θ_2 at the boundary and the temperature of the lower layer, and the heat exchange between the layers is proportional to the temperature difference between the layers. The expansion for θ_2 is assumed to be the same as for θ in Eqs.(3.53). For the description of motion in an annulus with fixed temperatures at both cylinders, the absence of any heat exchange at the free upper surface was supposed (in Chapter 4 the model will be generalized for the case of a vertical temperature gradient on the cylinder surfaces). If we denote the friction coefficients at the lower surface and between the layers as $2k_0$ and k_0' (non-dimensionalized by the Coriolis parameter f_c) and similar non-dimensional heat exchange coefficients as $2h_0$ and h'_0, then in the r.h.s. of the spectral equation system (3.54)-(3.57), the following terms will be added:

$$\text{to} \quad \frac{\partial \psi_i}{\partial t} = ... - k_0(\psi_i - \tau_i) \tag{3.59}$$

$$\text{to} \quad \frac{\partial \tau_i}{\partial t} = ... - k_0\psi_i - (k_0 + 2k'_0)\tau_i \tag{3.60}$$

$$\text{to} \quad \frac{\partial \theta_i}{\partial t} = ... - h_0(\theta_i - \sigma_i) + h'_0\theta_{2i} \tag{3.61}$$

$$\text{to} \quad \frac{\partial \sigma_i}{\partial t} = ... + h_0\theta_0 - (h_0 - 2h'_0)\sigma_0 + h_0\theta_{20} \tag{3.62}$$

As in all geostrophic and quasi-geostrophic models, here the acoustic and gravity waves are filtered out. The transfer of angular momentum by the vertical motions and by the divergent part of horizontal motions is absent. A weak point of the model is the introduction of the phenomenological dissipation coefficients which have not yet been related to the real molecular coefficients of a fluid.

To completely determine the equation system (3.54)-(3.58), one should know the coefficients a_i and c_{ijk} which are determined by the choice of the system of the orthogonal functions F_i. The choice depends on the geometry of the region in study. For the cylindrical vessels, the natural orthogonal functions system are the Fourier-Bessel's functions and the expansion is of the form:

$$G = G_{00} + \sum_{n=1}^{\infty} G_{0n}F_{0n} + \sum_{m,n=1}^{\infty} (G_{mn}F_{mn} + G'_{mn}F'_{mn}) \tag{3.63}$$

where

$$F_{0n} = J_0^{-1}(j_{in})\, J_0(j_{in}r_0) \tag{3.64}$$

$$F_{mn} = \sqrt{2}\, J_{m-1}^{-1}(j_{mn})\, J_m(j_{mn}r_0)\, \cos m\varphi \tag{3.65}$$

$$F'_{mn} = \sqrt{2}\, J_{m-1}^{-1}(j_{mn})\, J_m(j_{mn}r_0)\, \sin m\varphi \tag{3.66}$$

and $r_0 = \frac{r}{b}$, r and φ are the polar coordinates, J_m is the Bessel function of the order m and J_{mn} is the n-th positive root of the equation $J_m = 0$. Coefficients in the expansions are chosen from the condition that the mean square of the function F inside the circle $r \leq 1$ is equal to unity.

A still greater simplification is the consideration of motions in a rectangular region which is equivalent to negligence of the vessel curvature: $2(b-a) << (b+a)$. Then the Cartesian coordinates are introduced (x and y) and an analysis is carried out in a channel confined by $y = 0$ and $y = \pi L$. The corresponding series of the orthogonal functions is represented by

$$F_{00} = G_{00} = 1 \tag{3.67}$$

$$F_{om} = \sqrt{2}\,\cos ny_0 \tag{3.68}$$

$$F_{nm} = \sqrt{2}\,\sin ny_0\,\cos mx_0 \tag{3.69}$$

$$F'_{nm} = \sqrt{2}\,\sin ny_0\,\sin mx_0 \tag{3.70}$$

where $x_0 = \frac{x}{L}$, $y_0 = \frac{y}{L}$. After calculating the coefficients a_i and c_{ijk} using expansion (3.63)-(3.66) or (3.67)-(3.70), the equation system (3.54)-(3.58) is solved which is added by the dissipative terms (3.59)-(3.62). We should note that in the absence of dissipation, the system (3.54)-(3.58) has three integrals of motion similar to the mean square of the temperature fluctuations, to the total kinetic energy (within the accuracy of an additive constant) and to the sum of the available potential and kinetic energy.

The further analysis of the spectral form of the equations depends on the number of the functions F_i which enter the expansion, i.e. on the choice of the combinations m and n. The choice of $n = 1$ and one value of m carried out by Lorentz (1962) is the simplest way to consider the stability of the axisymmetric regime and for an analysis of regions of stationary wave regimes with a given wave number m. Under the assumption that an axisymmetric flow is possible for all values of the external parameters the stability of the flow is considered in dependence on the analog of the thermal Rossby number θ^*_A, the externally prescribed temperature expansion coefficient θ_{20} and an analog of the Taylor number, i.e. the value k^{-2}: for the simplicity it is assumed that $k' = h' = \frac{h}{2} = \frac{k}{2}$ which is a natural consequence of taking into account the dissipation through the coefficient k non-dimensionalized by the Coriolis parameter, or by a value proportional to the angular velocity Ω. The characteristic stability diagram for stationary wave regimes with different wave numbers m is presented in Fig.58, where the digits show the corresponding values of m and the regions are hatched in which the wave regimes are stable with the neighboring wave numbers. Note that these results have been obtained by an analytical consideration of the model with small computations determining the expansion coefficients. A good qualitative agreement between the model and experimental results presented in Fig.47 is clear. The agreement concerns the possibility of the stationary regimes for different values of the wave number m and the hysteresis phenomenon.

Lorenz (1963) was able to obtain not only stationary wave regimes but also some vacillation and irregular ones for which he used $n = 1$ and 2 and one value of m, i.e. he considered the two mode interaction (actually, he considered the expansion (3.67)-(3.70) for the straight channel). It is evident that in the model with its small number of modes only the amplitude vacillation can be reproduced but not the structural vacillation when a large number of additional modes is exited. A good qualitative agreement of the model and experimental results is observed in the upper part of the diagram for axisymmetric and waves regimes. At small values of θ^*_A or the Rossby number Ro_T, and at large rotation velocities Ω, or the Taylor number Ta, or parameter k^{-2}, the model has only regular regimes while in the experiments there is a considerable growth of complexity of the motion which requires a large number of the modes.

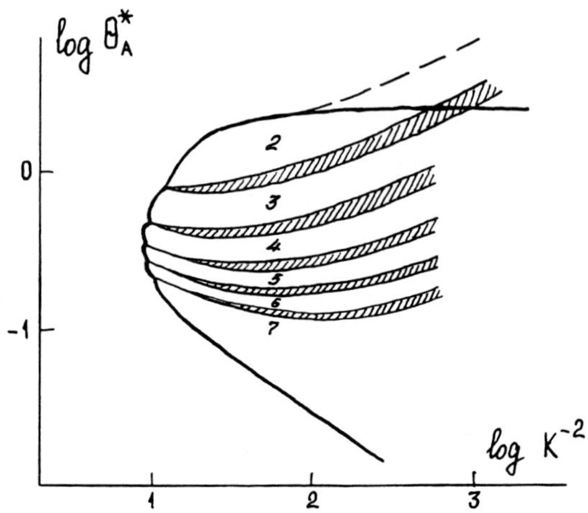

Figure 58: Criterion for stability of the axisymmetric regime (heavy curve), and criteria for transitions between wave numbers (thin curves), when all wave numbers are allowed. Heavy dashed curve indicates extreme values of θ_A^*, for corresponding values of k^{-2}, for which the wave regime can exist. Shaded areas indicate the existence of two wave numbers together. (After Lorenz, 1962, courtesy of J. Atmos. Sci).

An increase of the number of modes considered in the model changes only weakly the stability boundary of an axisymmetric flow but the increase allows one to analyze in more detail the region where the axisymmetric flow is unstable. The presence in the system of two modes is a necessary condition for the existence in the model of vacillation regimes. In dependence on the external parameters, the angular rotation rate and the temperature difference between the cylinders, both modes with $m = 1$ and $m = 2$ for a given n can develop separately (e.g. the wave with $m = 1$ is developing and the wave with $m = 2$ damping) or simultaneously. In the last case, the stationary waves of each mode propagate together without changing their forms: the so-called wave regime of the mixed mode. The non-linear character of the model (3.54)-(3.58) allows one to consider also an interaction of two different modes when there are ranges of the system's external parameters for which a development of one mode causes the development of the second mode. Such non-linear interactions lead to the appearance in the system of vacillation regimes and also of irregular geostrophic regimes.

An analysis of the two mode interactions to clarify the nature of vacillations is a general method for many non-linear models. Pedlosky (1977) used a two-layer dissipative model, considered non-linear wave interaction and found ranges of external parameters inside the region of the linear instability where there are vacillation regimes. For a model with continuous vertical stratification and dissipation in the Ekman layers at the lower and upper boundaries, the vacillation regimes were analyzed. Barcilon & Drasin (1984) considered a case on non-linear interaction of two modes with the same wave number but with different vertical structures propagating with different rates relative to the vessel walls. This case was analyzed to clarify the amplitude vacillation and was observed in the experiments.

3.8 Thermal and topography inhomogeneities

Convective motions in rotating annuli at homogeneous and constant temperatures at the cylinder walls model only the most general features of atmospheric circulation: the size of warm air masses near the equator and their subsidence in polar regions. Now, we consider the possibilities of experiments and simple theoretical models for a more detailed simulations of large scale atmospheric motions. Here, we shall consider stability problems for a zonal (axisymmetric) flow and stationary waves in the presence of thermal and topographic inhomogeneities at the underlying surface. From the geostrophic point of view, such problems are of interest because the differences in temperature and height between continents and oceans substantially transform large scale atmospheric motions; besides, large mountain ranges can lead to the formation of atmospheric blocking when at a certain place the circulation does not change during a considerable time.

From the mathematical point of view the description of the influence of topography and thermal inhomogeneities are similar to each other as well as their influences on the large scale motions (see Charney & DeVore,1979, Charney & Straus,1980).

3.8.1 Thermal inhomogeneities

One incentive to consider the thermal inhomogeneities in the model of atmospheric general circulation was an analysis of the surface air temperature in the middle latitudes performed by Hartmann & Short (1979) and presented in Fig.59. An almost sinusoidal dependence, i.e. $T(\varphi) \propto \sin 2\varphi + const$, on the longitude φ is clear with two maxima and two minima corresponding to two continents and two oceans. Such a dependence is especially evident during the winter time and it is in the winter when the general circulation anomalies are especially conspicuous. This is caused by the fact that the temperature of the ocean surface is always positive in the temperate latitudes while it is decreasing as one moves on the continents further away from the oceans.

The influence of a thermal inhomogeneity can be traced without difficulties using the two layer geostrophic model. The setting of the problem differs from the one in sections 3.7 only in the fact that the temperature at the underlying surface is not constant any more but depends on the azimuth (longitude). Therefore, the spectral form of equations (3.54)-(3.58) does not change, as well as the added dissipative terms.

The difference appears at the choice of the expansion terms and is caused by the necessity to consider the wave motions with one wave number $m = 2$ corresponding to the temperature difference's proportionality to $\sin 2\varphi$. Therefore, in the expansions it is necessary to consider, at least, two different wave numbers m and one number n. Boubnov et al (1991) carried out such an analysis for $m = 2$ and 3 and $n = 1$ for a cylindrical annulus geometry, i.e. with the choice of the Bessel functions as the basic orthogonal functions (3.63)-(3.66). For the temperature at the bottom of the vessel, the expansion is

$$\theta^* = B_T L^2 f_c (\theta_0^* + \theta_A^* F_{01} + \theta_h^* F_{21}) \tag{3.71}$$

where

$$F_{01} = J_0^{-1}(j_{11}) J_0(j_{11} r_0) \tag{3.72}$$

$$F_{21} = \sqrt{2} J_1^{-1}(j_{21}) J_2(j_{21} r_0) \cos 2\varphi \tag{3.73}$$

while for the case of the homogeneous bottom temperature only F_{01} was present and the problem was determined by the only temperature parameter θ_A^*. Now the second parameter appears: the amplitude of the harmonic temperature change θ_h^*.

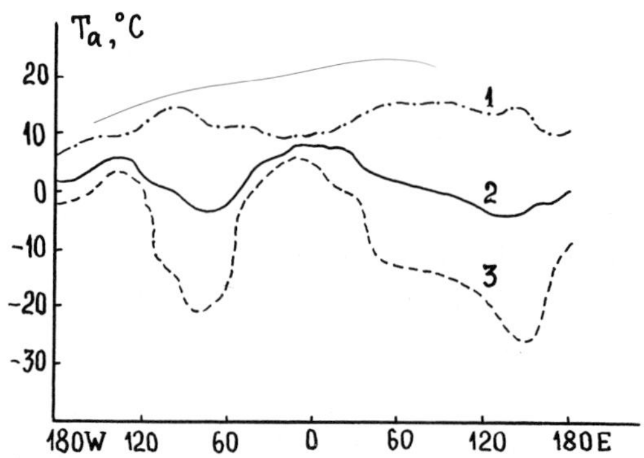

Figure 59: Mean surface air temperature T_a at the latitude 55° N: 1-for July; 2-mean annual temperature; 3-for January. (After Hartman & Short, 1979, courtesy of J.Atmos. Sci.).

From the mathematical point of view, the asymmetry of the bottom temperature inhibits the existence in the system of an axisymmetric flow which cannot satisfy the boundary condition at the lower boundary. Therefore, the axisymmetric regime is replaced by a new stationary regime corresponding to $m = 2$, but it is at rest relative to the vessel walls. The stability region of this regime differs only weakly from the region of the stability of the axisymmetric regime and also has an anvil shape. The boundary of the stability region of the stationary regime at an increase of θ_h^*

first shifts to the left at the regime diagram in coordinates of analogs of the thermal Rossby number θ_A^* and the Taylor number k^{-2}, and then in the upper part of the diagram shifts to the right and in the lower part to the left so that the stability curves at θ_h^* =0.1 and θ_h^* =1.0 are crossing each other.

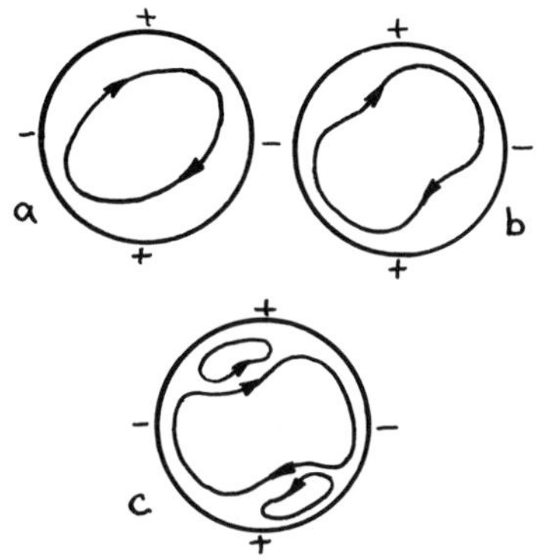

Figure 60: Stream function patterns of the stationary regime for the clockwise rotation of the vessel: $\theta_A^* = k = 0.6$, (a) $\theta_h^* = 0.2$, (b) $\theta_h^* = 0.4$, (c) $\theta_h^* = 1.0$. (After Boubnov et al, 1991, courtesy of Gordon and Breach Science Publishers).

The flow pattern in the stationary regime changes qualitatively with the changes of the external temperature amplitude θ_h^* at the bottom of the vessel. The stream function patterns for various values of θ_A^* is presented in Fig.61. A characteristic feature of the patterns is the position of its symmetry axis between the maxima and minima of the heating. A critical value of θ_h^* exists at which a qualitative transformation of the flow patterns takes place and in addition to the main motion with $m = 2$ two additional vortices appear.

In the region of the stationary flow instability for the basic mode with $m = 2$ at expansion over the given set of the orthogonal functions, the following regimes are possible as shown in Fig.61: regular waves with $m = 2$ and 3, propagating with the constant angular velocity relative the the vessel walls and a vacillation regime with $m = 3$. The vacillation regime differs from the corresponding regimes for homogeneous heating. The propagation of the wave with $m = 3$ relative to the cylindrical walls is changing with time and the vortices in the system differ strongly in size from each other. The stream function patterns for various moments of this auto-oscillating motion is presented in Fig.62. The main feature of such a motion is a sequential formation of one vortex larger then two others. This large vortex is delayed near one of the two temperature minima while the other two vortices moves towards the large one with different velocities. After some time, this large vortex starts to move from the site of the temperature minimum and to decrease in size while is moving away from the minimum. At the same time, the vortex moving before it starts to increase in size and is breaking down as it approaches to the second temperature minimum. Then, the third vortex start to increase in size approaching the first temperature minimum and so on.

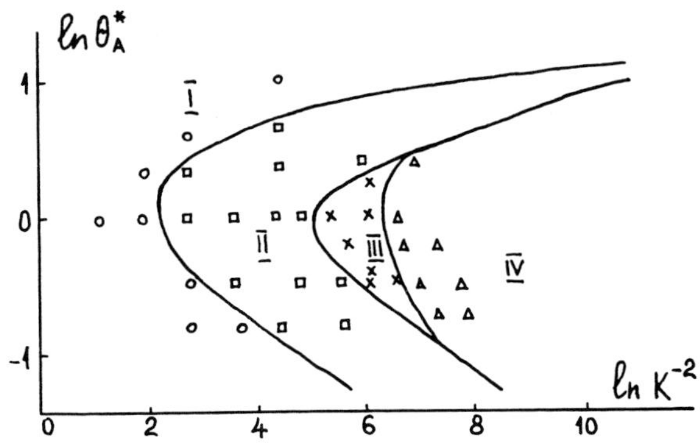

Figure 61: The stability diagram of regimes at $\theta_h^* = 1.0$. I - the region of the stationary regime, II - the wave regime with $n = 2$, III - the vacillation regime, IV - the wave regime with $n = 3$. (After Boubnov et al, 1991, courtesy of Gordon and Breach Science Publishers).

Boubnov et al (1988) studied this problem experimentally. To produce an azimuthally inhomogeneous heating corresponding to $m = 2$, two vertical heaters and two coolers were inserted into the external cylinder. As a result, almost sinusoidal temperature changes were produced at the surface of the external cylinder. The internal cylinder had a homogeneous temperature. For the amplitude of the external heating of 8K and for the temperature difference between the cylinders $\triangle T > 2$K, the experimental results on the regime diagram does not practically differ from the case of the homogeneous heating, i.e. the influence of the heating inhomogeneity is insignificant. As smaller values of $\triangle T < 2$K, new regimes appear: a wave with $m = 2$ resting relative to the vessel walls, a quasistationary wave with $m = 4$ which rests for a considerable time relative to the walls and then jumps to a new quasistationary position and a vacillation regime with $m = 3$, the changes in which are taking place precisely as it described by our simple model: a large vortex arises and delays near one temperature minimum which then moves away and another large vortex appears near another temperature minimum.

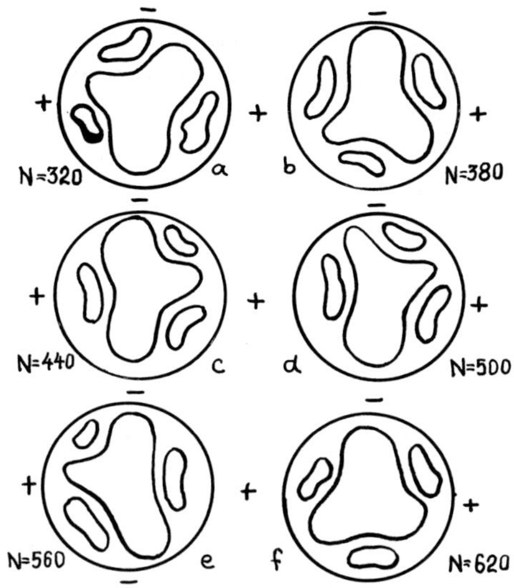

Figure 62: The pattern of the stream function for different times (N is the step of integration) at $\theta_h^* = 1.0$, $k = 0.05$, $\theta_A^* = 0.9$. (After Boubnov et al, 1991, courtesy of Gordon and Breach Science Publishers).

3.8.2 Topography inhomogeneities

From the theoretical point of view, the influence of the underlying surface inhomogeneities on the formation of the blocking situation in the atmosphere was considered by many authors. The analogy between thermal and topography inhomogeneities was shown at an example of the two layer barotropic model by Charney & DeVore (1979). Charney & Straus (1980) and Li et al (1986) considered a baroclinic model (3.54)-(3.58) added by inclusion of the surface inhomogeneity and by the variable Coriolis parameter for the two modes $m = 1$ and $m = 2$ and one value of n for the case of the plane geometry (3.67)-(3.70). In the expansion of the height of the underlying surface, only one term is taken into account

$$h = 2Hh_k \cos nx \, \sin y \tag{3.74}$$

The analysis of the model shows the existence of the equilibrium blocking situations caused by the topography wave (3.74). The instabilities of these equilibrium situations causes the formation of vacillation cycles in which the wave in the atmosphere propagates westwards with the period of 5 to 15 days and it interacts with the fluctuating zonal flow.

Experimental studies by Guoqing et al (1987) in rotating annuli were carried out for azimuthally homogeneous heating but with a wavy bottom: $h = h_0 \sin 2\varphi$, $0 \leq 2\varphi \leq 2\pi$, $h_0 = 1.75$ cm and the height of the fluid H =10 cm. Motions with and without topography were compared at the transitions from the axisymmetric regime to the amplitude vacillation and to geostrophic turbulence at changes of the thermal

Rossby number from 0.01 to 20. The results obtained are very similar to the results for the case of the thermal inhomogeneities. The region of axisymmetric motion at large temperature differences between the cylinders practically does not change; for the non-axisymmetric motions the influence of the topography starts to be felt: at $Ro_T > 2$, the wave is only slightly transformed, but at $0.06 \leq Ro_T \leq 0.2$, there was observed a strong amplitude vacillation in the shape of the waves of a varying form propagating relative to the vessel walls. Instead of the wave with a constant speed and a weak smooth change of its form as in the case of the homogeneous bottom, the wave motion arises with a periodical formation of intensive vortices in the sites of the bottom minima and with a subsequent delay and shift with transformations of the forms of vortices.

In the lower part of the regime diagram at small temperature differences $\Delta T < 2$K, a regular regime with $m = 4$ arises resting relative to the vessel walls and equivalent to the quasistationary regime of Boubnov et al (1988) for the case of inhomogeneous heating. The difference between experiments with topography and thermal inhomogeneities in some details (absence of the standing wave with $m = 2$, in types of vacillations) is determined to a large degree by geometry differences in the two installations: in the case of topography, the emphasis is shifted toward motions with larger wave numbers.

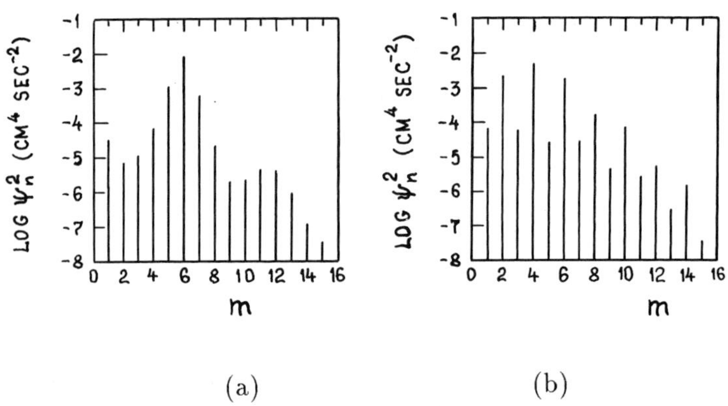

Figure 63: Time mean zonal wave number spectra near the midradius for experiments with: (a) no mountains; (b) mountains. (After Li et al,1986, courtesy of J. Atmos. Sci.).

An analysis of the motion spectra was carried out by Li et al (1986) with regard to external conditions for which in the absence of the topography (Fig.63a), the dominant wave number $m = 6$. The Fourier analysis has been performed for the stream function at each time moment at 16 radial cross-sections. Beside the main frequency $m = 6$, the spectrum has prominent maxima at $m = 1$, $m = 11$ and 12 corresponding to an interaction of the basis frequency with the next energy-rich mode $m = 5$.

For the experiments with the topography corresponding to $m = 2$, the spectrum changes (see Fig.63b) and the maximum of the spectrum is at $m = 4$, the smaller maxima are at $m = 2$ and at other even number modes whose energy exceeds considerably the energy of odd number modes. The dominant mode $m = 6$ in the presence of the topography has about half of the energy of the no-topography case while the other energy-rich modes $m = 4$ and 2 increase several times their energy in the presence of topography as compared to the homogeneous bottom case.

3.9 Specific heating cases

In the preceding sections of Chapter 3, the main attention was paid to studies of convective motions in rotating annuli with a stationary heated external cylinder and a cooled internal one. Not dealing with the influence of the vertical stratifications which will be described in the following Chapter 4, let us consider some other cases of heating: internal one, periodically changing in time, the hysteresis phenomenon. These cases are important for a fuller understanding of the mechanisms of the baroclinic instability in fluids.

3.9.1 Internal heating

Heat sources within the fluid were usually produced by transfer of electric current through electroconducting fluids with a given ohmic resistance (see Hide & Mason,1970,1975, Tanaka & Ukaji,1986). Such experiments were carried out either to analyze the influence of various conditions of heating on the flow patterns of the arising wave regimes, or to study convective motions in annuli with a small internal radius.

Examples of different cases of external and internal heating are presented in Fig.64 taken from Hide & Mason (1970,1975), where temperature profiles and radial temperature gradients are presented schematically. The last quantity for the axisymmetric flow is proportional to the zonal flow at the upper surface due to the wind equation. Also shown are the flow patterns at the upper surface in the stationary wave regime. Considering the heat flux through the side walls taking account of the thermal wind equation and ageostrophic motion in the Ekman layers, it can be shown (see Hide & Mason,1970) that the total heat flux H^* through the cylinder boundaries should satisfy the conditions $H^* \propto \oint v^* r \, d\varphi$ where v^* is the zonal (azimuthal) component of the quasi-geostrophic motion at the surface near the boundaries. In the case when the heat flux through the boundary is zero, the velocity zonal component should also vanish. In Fig.64 the basis rotation of the system is anticyclonic; therefore, in the case 64a when the heating is at the external cylinder and the cooling at the internal one we have a conspicuous jet stream without additional vortices which coincides in its direction with the general system rotation and the wave propagates relative to the vessel walls in the same direction. In case 64b, when the internal cylinder is heated and the external one is cooled, the motion in the wave is in the direction opposite to the system rotation and the wave retards from the walls.

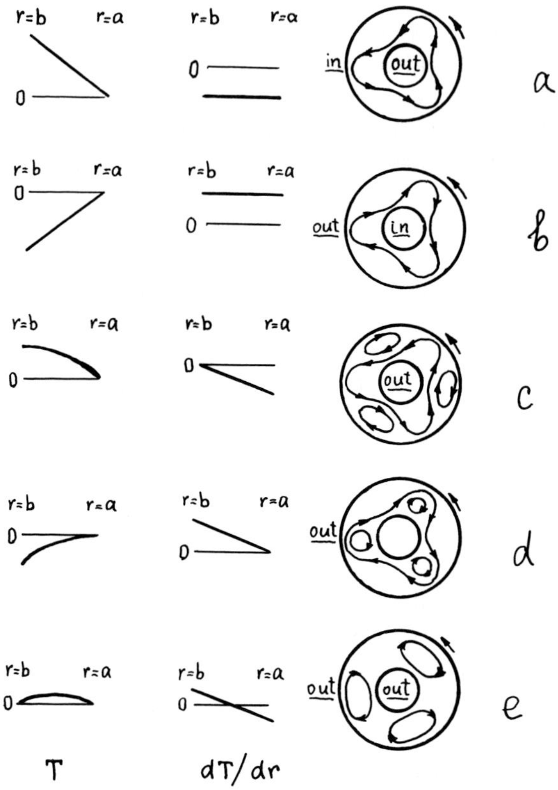

Figure 64: Schematic illustration of the variations of the impressed temperature T and $\frac{dT}{dr}$ with radius r (basic rotation anticlockwise). First row (a): outer wall heated and inner wall cooled: top-surface flow is generally cyclonic. Second row (b): inner wall heated and outer wall cooled: top-surface flow is generally anticyclonic. Third row (c): internal heating, inner wall cooled: as no heat the outer wall in symmetric regime and the presence of closed anticyclonic in contact with the outer wall in the wave regime. Fourth row(d): internal heating, outer wall cooled: as no heat crosses the inner wall in symmetric regime and the presence of closed anticyclones in contact with the inner wall in the wave regime. Fifth row (e): internal heating, both walls cooled: a wave regime flow pattern consisting of closed anticyclones in contact with both inner and outer walls. (after Hide & Mason, 1975, courtesy of Adv. Phys.).

In case 64c, the internal cylinder is cooled and the heating is within the fluid. Here the heat flux is absent at the external cylinder and near it anticyclonic vortices arise in order to satisfy the condition $v^* = 0$. At the opposite case, i.e. of the external cylinder cooling and internal heating (Fig.64d), the vortices arise near the internal cylinder and the jet stream changes its direction of motion. If both cylinders are cooled and there is an inside heating (Fig.64e), the mean zonal velocity is negative near the external cylinder and positive near the internal one, which leads to breaking down the jet stream and to the formation of vortices occupying all the space between the cylinders.These examples underline the necessity of a cautious and critical approach for experiment's in the annuli. In particular, any asymmetry in the boundary conditions due to interaction with the environment may cause the appearance of mixed motion variants.

An example of a specially designed mixed variant of heating are the experiments by Tanaki & Ukaji (1986) where the wave with $m = 4$ immovable relative to the walls was modelled by generating a constant temperature difference at the side walls and by an additional heating within the fluid. Two oppositely acting mechanisms lead to the formation of standing stationary waves similar to the waves which do not propagate relative to the vessel walls due to thermal or topography inhomogeneity as has been described in section 3.8.

To check the regularity of the wave motions in rotating annuli with a very small internal cylinder radius $a << b$, or $\delta^* = \frac{a}{b} << 1$, a vessel was used in which as an internal cylinder a wire was placed with an electric current and the outer cylindrical wall was cooled (see Hide & Mason,1970). It was found that the regular motions do exist in the system. In such a system, according to Eqs.(3.34) the wave motions can exist only with $m = 2$ which was confirmed in the experiments by Hide & Mason (1970) and Boubnov (1984): motions with $m = 1$ and 2 have been observed and the axisymmetric circulations. The observed wave motions have a distorted wave shape and the degree of distortion increased with the decreasing value of δ^*.

The problem of the motion's regularity at small values of δ^* was considered theoretically by Boubnov (1977) for the two-layer geostrophic model described by Eqs.(3.54)-(3.58) with dissipative terms (3.59)-(3.62). The expansion on the orthogonal Bessel functions was used but not for the cylinder with the zero internal radius $a = 0$, as in Eqs. (3.64)-(3.66) but the eigen functions for the region $a \geq r \geq b$ were used:

$$F_{0n} = [Y_1(a_{1n})\, J_0(a_{1n}\, r_0) - Y_0(a_{1n}\, r_0)\, J_1(a_{1n})]\, \frac{1}{\alpha_n} \tag{3.75}$$

$$F_{mn} = [Y_m(a_{mn})\, J_m(a_{mn} \cdot r_0) - Y_m(a_{mn}\, r_0)\, J_m(a_{mn})]\, \frac{\cos m\varphi}{\beta_{mn}} \tag{3.76}$$

$$F'_{mn} = [Y_m(a_{mn})\, J_m(a_{mn}\, r_0) - Y_m(a_{mn}\, r_0)\, J_m(a_{mn})]\, \frac{\sin m\varphi}{\beta_{mn}} \tag{3.77}$$

where J_m is the Bessel function of the first kind and of the m-th order, Y_m is the Bessel function of the second kind of the m-th order, a_{mn} is the m-th positive root of the equation

$$Y_m(a_{mn})\,J_m(a_{mn}\,\delta^*) - J_{mn}(a_{mn})\,Y_{mn}(a_{mn}\,\delta^*) = 0 \tag{3.78}$$

and the coefficients α_n and β_{mn} are chosen from the conditions that the mean square of the functions F in the region $a \le r \le b$ is equal to unity.

The two-mode expansion with $n = 1$ and 2 and one value of m was used. It was found that the introduction of the new expansion caused new non-linear interactions due to the geometry of the system, complicating the flow patterns. These complications are substantial for $\delta^* < 0.2$, but for larger values of δ^*, there was no qualitative difference between motions for small and large values of the internal cylinder radius.

3.9.2 Periodic changes of the external temperature

Experiments with cyclic changes of the temperature $\triangle T$ between the rotating cylinders (see Buzina et al,1978 and Snyder & Youtz,1969) were carried out in the region of the regular wave motions. In the last work, a transition from the motion with $m = 2$ to one with $m = 3$ was studied in the system with a weak hysteresis. It was found that at the cyclic temperature change with the amplitude about one third of the $\triangle T$ value at the slow temperature changes, the transition from the motions with $m = 2$ to the one with $m = 3$ took place at about the average value of the temperature, as in the case of the stationary temperature conditions. At a sufficiently fast temperature change, the transition was substantially retarded and in the limit of a very fast temperature change it did not take place at all.

In the work by Buzina et al (1978), the choice of the system parameters was motivated by the wish to study the role of the seasonal temperature variations on the atmospheric motions. As the basic period of the temperature variations the period equal to 365.25 periods of the system rotation was chosen. The control experiments with no changes in the $\triangle T$ value were also carried out together with experiments on the studies of the hysteresis phenomenon at slow changes of the external parameters while approaching a certain point in the parameter space. These experiments have shown that the phenomena observed in any point of the dimensionless parameter space depended on the way by which the point had been reached and also on the rate of changes of the external parameters. However, in all cases the experiments revealed a deterministic fluid behavior in various points of the parameter space differing from the case of the cyclic changes of the external parameters where the fluid behaves non-deterministically. It was supposed that the cyclic experiments are in between the two deterministic limits: experiments at a constant temperature difference (equivalent to extremely fast cyclic changes) and experiments of the hysteresis studies (very slow cyclic changes). Therefore, the results are not rigorously dependent on the value and the rate of change of the external parameters, though they reveal some general statistical regularities in some regions of the parameter space. In dependence on the values of the external parameters, the cyclic forcing in these experiments leads either to transition from a vacillation regime with $m = 4$ to the stationary wave regime with $m = 5$, or, at slower periods, to the vacillation regime with a larger value of m , or to a vacillation regime with a wide spectrum of wave numbers. All these transitions are a natural consequence of the resonance of the eigen modes exited in the system with an external forcing frequency and of the subsequent mode interactions.

An example of the external periodic forcing of a simpler non-rotating convective system was experimentally investigated by Boubnov & Gledzer (1981). In this study, the temperature difference between the upper and lower part of the sphere was cyclically changed with a frequency ω^*. As a result, in the system resonant oscillations were exited. The spectrum of these oscillations had maxima at the forcing frequency ω^* as well as at the halved frequency $\frac{1}{2}\omega^*$. At some values of the external parameters, the maximum at the halved frequency was considerably larger than at the basic frequency ω^*.

3.9.3 Hysteresis

The phenomenon of hysteresis was noted at the first stage of the studies of the convection in rotating annuli when it was observed that there were points (or regions) in the external parameter space where the flow patterns depended on the way these points had been reached. The greatest attention to the hysteresis investigation was paid in the works by Fultz et al (1958), Kaiser (1970), Hide & Mason (1970). Besides the fact that the transition from one type of motion to another one takes place at different values of the external non-dimensional parameters one should also note that during the transition to a new state there are regimes where motions are excited with several neighboring wave numbers.

Lacher et al (1977) proposed a qualitative explanation of the hysteresis phenomenon and some other effects arising at an approach to a stationary state based on the theory of catastrophes. It was supposed that the total energy E of the system (or some components of the energy: available potential, kinetic) can depend only on the system rotation rate Ω, the temperature difference between the cylinders $\triangle T$ and time t. It was also supposed that this dependence is a smooth function of these variables for a fixed set of mutual conditions. It was also supposed that

$$\frac{dE}{dt} = f(\Omega, \triangle T, t) \tag{3.79}$$

and that the function $f(\Omega, \triangle T, E)$ is such that its value at a point in the parameter space $(\Omega, \triangle T, t)$ has a tendency to an increase or a decrease with time at this point. The equilibrium points of the Eq.(3.79) must lie at the surface

$$f(\Omega, \triangle T, E) = 0 \tag{3.80}$$

in the space $(\Omega, \triangle T, E)$. The stable equilibrium corresponds to energy stationary values. Further more, it was supposed that the model is "generic", or roughly speaking is prescribed by a potential function in the parameter space $(\Omega, \triangle T, E)$ and that the qualitative properties of the function do not change at small perturbations. The equilibrium surfaces for various regimes in the rotating annuli are presented in Fig.65.

This figure shows that it is principally important how a point is approached at changing the external parameters. E.g., if the approach is at increasing the thermal Rossby number Ro_T (or $\triangle T$) and a fixed Taylor number Ta (or Ω) then at reaching the critical values (internal edge of the folding in Fig.65) a jump transition to the above laying surface is taking place because the motion at the parameters on the internal part of the folding is unstable. At decreasing the value of Ro_T, the jump transition takes place at the external point of the surface inflection and as a result the jump is on the lower laying surface. The projection of three-dimensional space in Fig.65 is equivalent to a usual regime diagram and shows the presence of regions in which different regimes are possible in dependence on the way of their approach.

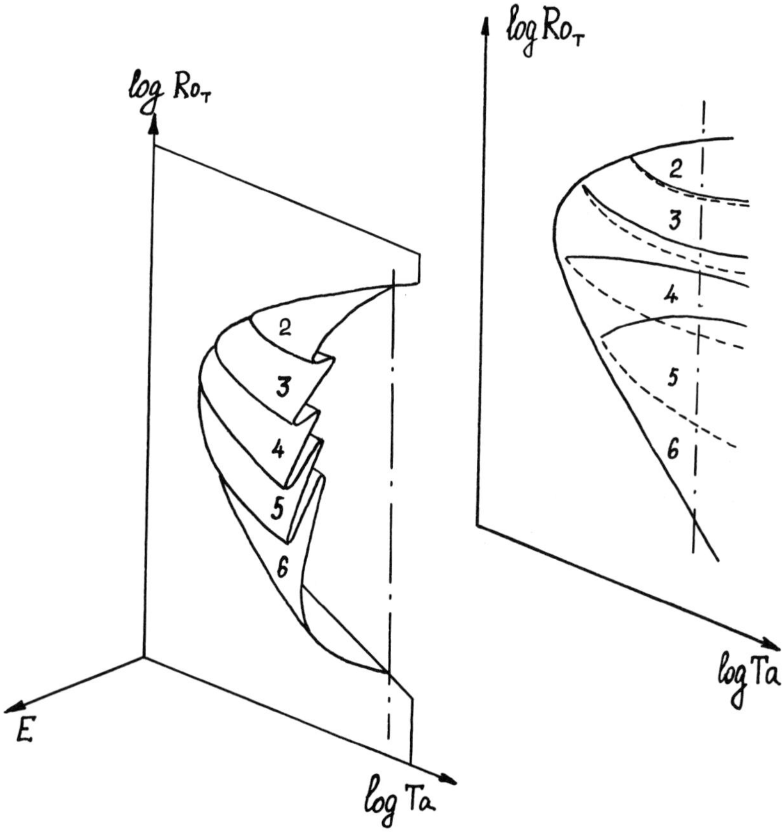

Figure 65: A schematic equilibrium surface represents the experimental results shown in Fig.58. The solid and dashed lines outlining the wave flow pattern constitute the catastrophe set of the model. The numbers on the surface indicate the wave pattern observed when E lies on this portion of the surface. E can represent the available eddy potential energy or the thermal variance of the flow field. (After Lacher et al, 1977, courtesy of Amer. Sci.).

This same model explains naturally why for a fast spinning-up of a vessel with a fluid and at the same final conditions, regimes arise in the system with various wave numbers and energy levels (see Hide & Mason,1970). This means that the energy E is not only a function of the final state of the system $(\Omega, \triangle T)$ but also of the way by which this final state was reached. However, we wish to stress that the process of hysteresis is basically differing from a more complicated non-deterministic behavior of the system in the case of the cyclic heating of the system.

Chapter 4
VERTICALLY AND HORIZONTALLY INHOMOGENEOUS HEATING

In the nature and in many technical applications there are quite often situations when there is an external heating not only in the horizontal but also in the vertical. From the theoretical point of view, this problem is equivalent to the one named after Eady, however the assumption of the constant vertical temperature gradient within the fluid does not agree well with the experimental results. Therefore, for comparison of these results with the theory one considers not only the external temperature gradient $\frac{\partial T_0}{\partial z}$ but also the local internal one $\frac{\partial T}{\partial z}$ formed in the experiments by convective motions and determined by the horizontal gradient $\frac{\partial T_0}{\partial r}$. Therefore, it is natural to continue considering convective motions in rotating fluids with a mixed heating. The simplest variant of heating will be that when the external temperature will be constant at the edge of the heated surfaces. Two limiting cases are possible here: first, modification of the problem of a plane rotating layer with inhomogeneously heated bottom when a jet stream arises near the heated horizontal surface, and, second, an inhomogeneously heated annulus with constant temperature gradient at the side walls. In the first case, the temperature vertical gradient inside the fluid changes strongly in the radial direction. In the second case, the radial change is not large, but the external temperature vertical gradient (which can be of both signs) influences strongly the motion patterns.

The criterium for transition from the first regime to the second one is the existence in most part of the fluid volume of an unstable stratification. This criterium determines the difference in the approach of these problems: in one case, the main attention is paid to the behavior in the boundary layers assuming that the main fluid core is almost homogeneous; in the second case, the main attention is paid to studies of the fluid core. Here, we consider sequentially these limiting cases, transition between them, some mixed cases of heating and also the diffusion circulation generated in the fluid at the density stratification which is similar in many respects to the case of baroclinic instability in thermally stratified fluids.

4.1 Temperature gradient at horizontal bottom

The problem of convective motions in a rotating annulus with a radial temperature gradient at its bottom was considered by Hignett et al (1981) at the condition of the thermal insulation of the side walls (with radii b and a) and upper rigid surface. The main difference from the cases of the Chapter 2 consists in the fact that in the presence of the horizontal temperature gradient within the fluid there is no critical Rayleigh number, i.e. motions exist at arbitrary small temperature differences. Three boundary layers exist: the thermal one, the dynamic one and the Ekman layer. In dependence of their relative thickness, the motion regimes change. The temperature distribution at the bottom, when temperatures at the outer and inner circle edges of the bottom are constant and differ by $\triangle T$, is equal to

$$T(r,0) = T(a) + \triangle T \frac{\ln \frac{r}{a}}{\ln \frac{b}{a}}$$

Following Hignett et al (1981), as non-dimensional system parameters we choose the numbers of Rayleigh Ra_D, Taylor Ta_D, Prandtl Pr and the geometry parameter $\delta_0 = \frac{h}{b-a}$. The characteristic length here is $D = b - a$. Therefore

$$Ra_D = \frac{\alpha g D^3 \triangle T}{\nu k}, \qquad Ta = \frac{4\Omega^2 D^4}{\nu^2} = Ek^{-2}. \tag{4.1}$$

These parameters differ from the usual ones in Chapter 1 and 2 in replacing the fluid height h by the horizontal distance D and by using the horizontal temperature difference $\triangle T$.

For the subsequent analysis, we shall make a very strong simplification: consider the stationary two-dimensional motion of an incompressible fluid, neglect the curvature of the vessel and assume the fluid density is a linear function of the fluid temperature. The largest difference in the assumptions from the reality is in the two-dimensionality of motions; however, it allows one to introduce the stream function ψ and to simplify strongly the subsequent analysis. Now, the equation system can be written as three equations for vorticity, transfer of heat and of azimuthal momentum:

$$\nu \nabla^4 \psi = \alpha g T' - 2\Omega v - J(\psi, \nabla^2 \psi) \tag{4.2}$$

$$k \nabla^2 T' = -J(\psi, T') \tag{4.3}$$

$$\nu \nabla^2 v = 2\Omega \psi_z + J(v, \psi) \tag{4.4}$$

where ψ is stream function: $u = \psi_z$, $w = -\psi_x$, v is azimuthal velocity component, T' is deviation of the temperature from its minimal value.

As for the thickness of the thermal boundary layer h_T: over the horizontal inhomogeneously heated surface we shall use the expression $h_T = D\, Ra_D^{-\frac{1}{5}}$ obtained by Rossby (1965) for large values of Ra_D in the case of no rotation. For the Ekman layer, we have $h_E = D\,Ta^{-\frac{1}{4}}$. It is natural to assume that the value of the ratio of the two thicknesses plays a determining role in the model so that we introduce the parameter

$$Q = \left(\frac{h_T}{h_E}\right)^2 = Ra_D^{-\frac{2}{5}}\,Ta^{\frac{1}{2}} \tag{4.5}$$

which is proportional to the angular velocity Ω.

For fixed sufficiently large values of the Rayleigh number and sufficiently large Prandtl numbers six flow regimes are determined by the relative thickness of the three boundary layers:

(i) $Q = 0$, no rotation

(ii) $Q << Pr^{-1} << 1$, very slow rotation

(iii) $Pr^{-1} << Q << 1$, slow rotation

(iv) $Q \approx 1$, moderate rotation

(v) $1 << Q << Ra_D^{\frac{4}{15}}$, fast rotation

(vi) $Q >> Ra_D^{\frac{4}{15}}$, very fast rotation

Now, we shall consider relationships between the various terms in the equation system (4.2)-(4.4) and the thickness of the boundary layers non-dimensionalized by the scale D.

(i) $\Omega = 0$. There are two boundary layers at the horizontal bottom: one is dynamically homogeneous with the (non-dimensional) thickness $Pr^{\frac{1}{2}}\,Ra_D^{-\frac{1}{5}}$ where the viscous force is balanced by the vorticity advection (first and fourth terms in Eq.(4.2)), the second is the thermal layer h_T (see above) where the viscous force is balanced by the buoyancy and the heat balance is fulfilled. At a homogeneous heating of a horizontal surface, the dynamical layer arises only after the onset of motions, i.e. for the values of the Rayleigh number larger than the critical one. The stream function in the main fluid core $\psi = O\left(k\,Pr^{\frac{1}{2}}\,Ra_D^{-\frac{1}{5}}\right)$; in the dynamical layer it tends to zero and in the thermal one to $k\,Ra_D^{\frac{1}{5}}$.

(ii) Very slow rotation: $Q << Pr^{-1}$. Azimuthal velocity $v = O(2\Omega D)$ and the Proudman-Taylor theorem is fulfilled within the fluid: $w = u = 0$. At the upper boundary, there are two layers: on of the thickness $D\,Ra_D^{-\frac{1}{5}}\,Pr\,Q^{-1}$ in which the value of the stream function decreases to $\nu\,Ra_D^{\frac{1}{5}}\,Q^{\frac{1}{2}}$ and a thinner Ekman layer in which ψ tends to zero. At the bottom surface, two previous layers exist in which v decreases in the dynamical layer due to advection and diffusion in the zonal flow and tends to zero in the thermal layer.

(iii) Slow rotation. Here, the scales of v and ψ become of order $k\,D^{-1}\,Ra_D^{\frac{2}{5}}$ and $k\,Ra_D^{\frac{1}{5}}\,Q^{-\frac{1}{2}}$. At the upper surface, there is the Ekman layer only. Inside the fluid, there is the homogeneous layer with $u = w = 0$. At the bottom, the dynamical layer is replaced by a thinner Ekman layer (because $Q << 1$) which is much thicker than the thermal boundary layer. The Ekman layer leads ψ and w to zero and $u = \psi_z$ is continuous between the Ekman and thermal boundary layers. The thickness of the last layer remains of order $Ra_D^{-\frac{1}{5}}$ which gives for the Nusselt number $Nu \propto Ra_D^{\frac{1}{5}}$.

(iv) Moderate rotation: $Q \simeq 1$. The Ekman and thermal layers at the bottom become of the same order due to the homogeneity of fluid in its main bulk and with the Ekman layer at the upper surface. At an increase of Q, or the system rotation rate Ω, the stream function decreases in the Ekman layer. At the same time, the thickness of the thermal boundary layer increases because the temperature vertical gradient decreases slower with an increase of Ω. As a consequence of this, the heat flux should decrease with Ω increasing. It follows from the thermal wind equation that the zonal velocity v should decrease with Q, while increasing for sufficiently large values of Q. Therefore, the value of v reaches its maximum at $Q \simeq 1$.

(v) For the fast rotation ($Ra_D^{\frac{4}{15}} >> Q >> 1$), the Ekman layer at the bottom becomes thinner than the thermal layer. The balance of forces in the main core and at the upper surface remains the same as in two previous cases. Due to this, the conditions in the upper and lower Ekman layers lead to values of ψ and v of order $k\, Ra_D^{\frac{1}{5}}\, Q^{-\frac{3}{4}}$ and $k\, D^{-1}\, Ra_D^{\frac{2}{5}}\, Q^{-\frac{1}{4}}$, respectively. The thickness of the thermal layer becomes $D\, Ra_D^{-\frac{1}{5}}\, Q^{-\frac{3}{4}}$ and the heat flux $Nu \propto Ra_D^{\frac{1}{5}}\, Q^{-\frac{3}{4}} = Ra_D^{\frac{1}{5}}\, Ta^{-\frac{3}{8}}$.

(vi) At very fast rotation Q increases to a much larger value than $Ra_D^{\frac{4}{15}}$. Here the thermal layer occupies the whole fluid layer. The fluid inside the layer becomes inhomogeneous, the values of ψ and v in both Ekman layers are of order $k\, Ra_D^{\frac{2}{5}}\, Q^{-\frac{3}{2}}$ and $k\, D^{-1}\, Ra_D^{\frac{3}{5}}\, Q^{-1}$, the value of the Nusselt number $Nu \to 1$, which corresponds to the regime of a pure heat conduction.

Experiments by Hignett et al (1981) were carried out with two fluids: paraffine and water, the maximal Rayleigh number was $7.5 \cdot 10^8$, the value of Q was in the range 0.3 - 10 which includes cases from slow to fast rotation. Baroclinic instability starts at $Q_{cr} = 3.4$ and then the flow characteristics change sharply. The value of Q_{cr} depends strongly on the geometry parameters of the system, Prandl number, etc.

Let us consider the main results of the experiments. The temperature field structure shows that outside the boundary layers the temperature decrease is linear with height and in the boundary layers its course changes from linear near the maximal temperature through a constant value to a decrease with a substantial negative gradient, corresponding to descending flow. Then, non-dimensional thickness of the thermal layer $h^* = h(b-a)^{-1}\, Ra_D^{-\frac{1}{5}}$ increases first rather slowly but starting with $Q = 1$ the rate of its growth increases much faster.

Measurements of the flow field showed that the ascending motion at the outer cylinder occupies no more than 9 per cent of the gap $D = b - a$ at a height of a few millimeters. The descending motion at the same time occupies about a third of the gap.

For any values of Q there are regions in the vessel where the fluid statistically unstable, i.e. $\frac{dT}{dz} < 0$. For small values of Q, these regions occupy a small part of the fluid, but for the fast rotation $Q > Q_{cr}$ the regions occupy more than a half of the fluid volume which leads to a transition from a stable flow to the development of wave perturbations whose nature was considered in Chapter 3. In the paper by Hignett et al (1981), for $Q > Q_{cr}$ it was suggested to divide the fluid into two layers: the lower one of thickness H^* near the bottom where the thermal wind is observed and the velocity shear $\frac{U}{H^*}$ arises and there is a stratification determined by the Vaisala-Brunt frequency N, and the upper layer with the constant velocity U and constant buoyancy with frequency N_U. Then the new parameters arise equivalent to thermal Rossby number Ro_T and to the static stability parameter S_z determining the rate of the vertical stratification in the problem of baroclinic instability. The role of the parameter S_z will be considered below in the section 4.2.

4.2 Stable vertical stratification in annuli

Because any small horizontal temperature inhomogeneity leads to the appearance of motions in the rotating annuli at any value of the thermal Rossby number within the fluid, the motions induce a vertical temperature stratification. Sometimes, as it was shown in Chapter 3, it may be considerably larger than the horizontal one : e.g. for an axisymmetric regime (see section 4.1) the mean vertical temperature is twice the value of the horizontal one.

Here, we consider the influence of an external temperature stratification on the convective motions in a annulus. For a theoretical consideration we shall use the two-layer geostrophic model but with different then in section 3.7 the upper boundary condition: instead of the thermal insulation we consider the constant temperature θ_1. The lower boundary condition remains the same: a constant temperature θ_2. Then the spectral form of the equations of motion (3.54)-(3.58) will remain the same, as well as the terms determining the friction (3.59)-(3.60) and only terms determining the heat exchange will change and instead of Eqs. (3.61)-(3.62), we shall have

$$\frac{d\theta}{dt} = ... + h_{00}\theta_{1i} + h_0\theta_{2i} - (h_0 - h_{00})\theta_i + (h_0 - h_{00})\sigma_i \tag{4.6}$$

$$\frac{d\sigma_0}{dt} = ... + h_{00}\theta_{10} + (h_0 - h_{00})\theta_0 - (h_{00} + 2h_0' + h_0)\sigma_0 - h_0\theta_{20} \tag{4.7}$$

where h_{00} is the heat exchange coefficient at the upper surface introduced in the same way as coefficient h_0 at the lower boundary in section 4.7.

A new external parameter can be introduced here: $\delta T = \theta_2 - \theta_1$, i.e. the temperature difference between the upper and the lower boundaries of the fluid. It is added to the horizontal temperature difference θ_A^* and the system angular rotation rate k^{-1}.

This example will demonstrate us some possibilities of a two-layer geostrophic model. Here we use the cylindric geometry and will restrict ourselves with $n = 1$ and one value of m, the azimuthal wave number. The expansion will involve only three functions: $F_A = F_{01}$, $F_K = F_{m1}$, $F_L = F'_{m1}$. As previously, for simplicity we assume $h'_0 = k'_0 = \frac{1}{2}k_0 = \frac{1}{2}h_0 = \frac{1}{2}h_{00}$. In this case, Eqs.(3.54)-(3.58) with dissipative terms will have the form:

$$\frac{d\psi_A}{dt} = -k_0\left(\psi_A - \theta_A\right) \tag{4.8}$$

$$\frac{d\psi_K}{dt} = -\alpha_m\beta_m\left(\psi_L\psi_A + \theta_L\theta_A\right) - k_0\left(\psi_K - \theta_K\right) \tag{4.9}$$

$$\frac{d\psi_L}{dt} = -\alpha_m\beta_m\left(\psi_A\psi_K + \theta_K\theta_A\right) - k_0\left(\psi_L - \theta_L\right) \tag{4.10}$$

$$\frac{d\theta_A}{dt} = -\omega_A + k_0\psi_A - 2k_0\theta_A \tag{4.11}$$

$$\frac{d\theta_K}{dt} = \alpha_m\beta_m\left(\theta_L\psi_A + \psi_L\theta_A\right) - \left(1 - \beta_m\right)\omega_K + k_0\psi_K - 2k_0\theta_K \tag{4.12}$$

$$\frac{d\theta_L}{dt} = \alpha_m\beta_m\left(\theta_A\psi_K + \psi_A\theta_K\right) - \left(1 - \beta_m\right)\omega_L + k_0\psi_L - 2k_0\theta_L \tag{4.13}$$

$$\frac{d\theta_A}{dt} = -\alpha_m\left(\theta_K\psi_L - \psi_K\theta_L\right) + \sigma_0\omega_A + k_0\left(\theta_{2A} + \theta_{1A} - 2\theta_A\right) \tag{4.14}$$

$$\frac{d\theta_K}{dt} = -\alpha_m\left(\theta_A\psi_L - \psi_A\theta_L\right) + \sigma_0\omega_K - 2k_0\theta_K \tag{4.15}$$

$$\frac{d\theta_L}{dt} = -\alpha_m\left(\theta_A\psi_K - \psi_A\theta_K\right) + \sigma_0\omega_L - 2k_0\theta_L \tag{4.16}$$

$$\frac{d\theta_0}{dt} = 2k_0(\theta_{20} - \theta_{10} - 2\theta_0) \tag{4.17}$$

$$\frac{d\sigma_0}{dt} = -\theta_A\omega_A - \theta_K\omega_K - \theta_L\omega_L + k_0(\theta_{10} - \theta_{20} - 3\sigma_0) \tag{4.18}$$

where the indices A, K, L correspond to the coefficients at the expansion of the functions F_A, F_K, F_L and

$$\alpha_m = \gamma_m\, j_{11}, \qquad \beta_m = 1 - \left(\frac{j_{11}}{j_{m1}}\right)^2,$$
$$\gamma_m = -2m\, j_{11} \int_0^1 \left[\frac{J_1(j_{11}r)\, J^2_{m1}(r)}{J_0(j_{11})\, J^2_{m1}(j_{m1})}\right] dr \tag{4.19}$$

The variables $\omega_A, \omega_K, \omega_L$ can be excluded from Eqs.(4.8)-(4.18) and we shall have a system of eight ordinary differential equations of the first order for the eight independent variables. Here it is supposed that the distributions of the variables at the upper and lower boundaries do not depend on the azimuth. Then all the spectral amplitudes with the indices K and L vanish and in the expansions for the temperatures only the terms with θ_{1A} and θ_{2A} remain, which are both positive at the cold internal and warm external cylinders.

Let us consider the simplest motion regimes. The zonal axisymmetric circulation corresponds to a circular motion for which all expansion amplitudes with indices K and L are equal to zero and in a stationary case all time derivatives also vanish. Then the stationary axisymmetric circulation is described by the following equation system:

$$\psi_A = \theta_A \tag{4.20}$$

$$\omega_A = -k_0\theta_A \tag{4.21}$$

$$\theta_0 = \frac{\theta_{10} + \theta_{20}}{2} \tag{4.22}$$

$$\sigma_0 = \frac{\theta_A^2 + \theta_{10} - \theta_{20}}{3} \tag{4.23}$$

where θ_A is the only real root of the equation

$$\theta_A^3 + \theta_A(\theta_{10} - \theta_{20} + 6) = 3\,(\theta_{1A} + \theta_{2A}) \tag{4.24}$$

which can be obtained from the rest of the equations (4.20)-(4.23) describing the stationary zonal circulation. Taking into account that $\theta_1 - \theta_2 = \theta_{10} - \theta_{20} = \theta_{1A} - \theta_{2A} = \delta T$, Eqs.(4.22)-(4.24) can be rewritten in the form:

$$\theta_0 = \theta_{10} + \frac{\delta T}{2}$$

$$\sigma_0 = \frac{\theta_A^2 + \delta T}{3} \tag{4.25}$$

$$\theta_A^3 + \theta_A\,(6 + \delta T) = 6\theta_{2A}$$

The transition from this model to the one by Lorenz (1963) takes place if $\theta_1 = \theta + \sigma$, i.e. when the heat transfer is absent at the upper boundary.

An analysis of the solution shows that the zonal flow is absent in the lower layer. The mean temperature is determined only by the boundary conditions according to (4.25). The static stability within the fluid, determined by the value of σ_0 (we remind the reader that the temperature difference between the layers is equal to 2σ) at the stable temperature stratification ($\delta T > 0$) can not be less than $\frac{\delta T}{3}$ for any value of the horizontal temperature difference determined by the value θ_{2A}. At the unstable temperature stratification ($\delta T < 0$), there is such a value of θ_{2A} increasing with $|\Delta T|$ for which $\sigma_0 \leq 0$. In this case at, a fixed value of the horizontal temperature difference ΔT an increase of the value of the vertical temperature difference leads to the formation of a heavier upper layer and the layers are mixed up. These effects cannot be taken into account in the framework of a given model which is applicable only for a statically stable system when $\sigma_0 > 0$.

Now consider the problem of the stability of the stationary zonal circulation relative to small azimuthal perturbations. The critical stability curve is determined by the equation

$$\frac{1}{k_0} - \frac{G_m(\sigma_0)}{\alpha_m \theta_A} = 0 \tag{4.26}$$

where for the system of Eqs.(4.8)-(4.18)

$$G_m^2(\sigma_0) = \frac{9(s+1)(s+2)}{[-5\beta_m\ s^2 + (3-7\beta_m)\ s + (12-2\beta_m)]\ \beta_m} \tag{4.27}$$

If we consider the critical stability curves for various δT, then at increasing the vertical stratification, i.e. δT, a shift of the wave regime region occurs to the right and to the lower part of the regime diagram in coordinates $\ln\theta_{2A}$ and $\ln k_0^{-2}$ (corresponding to the diagram in $(\ln Ro_T, \ln Ta)$ space). This is shown in Fig.66 for the case $m = 2$ at which for the unstable vertical stratification the curves end at $\sigma_0 = 0$. Note that the slope of the stability curves in the lower right part of the diagram for any value of m is always -1/2.

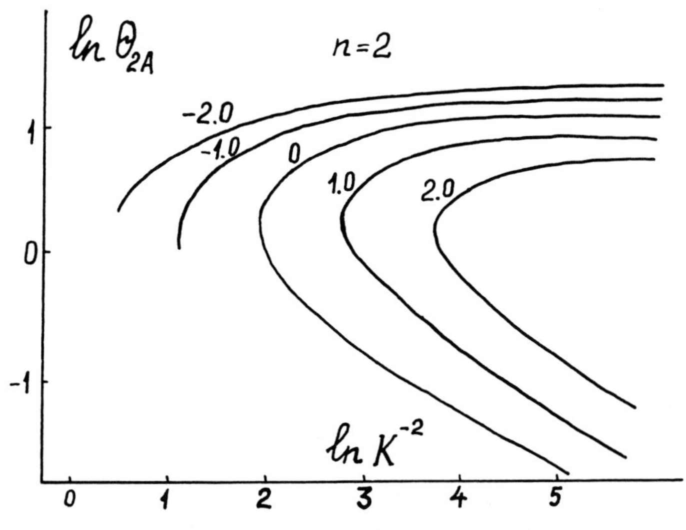

Figure 66: Regions of stability of zonal axisymmetric flow (to the left on the transition curves) for various values of the non-dimensional vertical temperature difference δT shown on the curves, where θ_{2A} is the amplitude of the external temperature difference ($k^{-1} \propto \Omega$) for the second mode. (After Boubnov & Golitsyn, 1988, courtesy of Gordon and Beach Science Publishers).

For the wave regime the solution is written as:

$$\psi_K = A_1 \cos \left[\frac{(5s+2)\ \alpha_m\, \beta_m\, \theta_A\, t_0}{3s+3} \right] \tag{4.28}$$

$$\psi_L = A_1 \sin \left[\frac{(5s+2)\ \alpha_m\, \beta_m\, \theta_A\, t_0}{3s+3} \right] \tag{4.29}$$

$$\theta_K = p_1\, \psi_K + q_1\, \psi_L, \qquad \theta_L = p_1\, \psi_L - q_1\, \psi_K \tag{4.30}$$

$$p_1 = \frac{1}{1+\beta_m^2\, G_m^2} \left[1 + \frac{(2s-1)\ \beta_m^2 G_m^2}{3s+3} \right] \tag{4.31}$$

$$q_1 = \frac{1}{1+\beta_m^2\, G_m^2} \left[\frac{(s+4)\ \beta_m^2 G_m^2}{3s+3} \right] \tag{4.32}$$

The relationship between the wave amplitude A_1 and the parameters of the zonal flow and the external parameters is determined by the equations

$$2\,\theta_A\,(\theta_{2A} - \theta_A) + 2\,(p_1^2 + q_1^2)\,A_1^2 = \sigma_0\,(\delta T - 3\sigma_0) \tag{4.33}$$

$$2\,k_0\,(\theta_{2A} - \theta_A) + \alpha_m\, q_1\, A_1^2 = -k_0\,\theta_A\,\sigma_0 \tag{4.34}$$

from which it follows that

$$A_1^2 = \frac{\sigma_0\,\theta_A^2\,(P^* - 1)}{q_1\, G_m\,(1 - V_1^*)} \tag{4.35}$$

$$\theta_{2A} = \theta_A \left[1 + \frac{\sigma_0\,(P^* - V_1^*)}{2(1 - V_1^*)} \right] \tag{4.36}$$

where $V_1^* = \frac{2(p_1^2+q_1^2)}{q_1 G_m}$ is the function of only σ_0 and

$$P^* = \frac{(3\sigma_0 - \delta T)}{\theta_A^2} \tag{4.37}$$

It can be shown that in the wave regime for all values of σ_0 the value of $V_1^* < 1$ and therefore $P^* > 1$ due to (4.35) and $A_1^2 > 0$. For the axisymmetric regime $P^* = 1$.

An analysis of Eqs.(4.34)-(4.37) for various values of m shows that with the vertical stratification parameter δT increasing in the system, motions can develop with smaller and smaller wave numbers, i.e. the vertical stratification acts as a filter for wave motions with large values of m. In other words, the stratification filters our higher modes. For example, if at $\delta T = 0.4$ the wave motions may exist with $2 \leq m \leq 7$ (a possibility of existence of such motions does not yet imply their stability), at $\delta T = 1$ the maximal m is 4 and at $\delta T = 3$ it is only 2.

The diagram for the existence of various regimes at $\delta T = 0.4$ is shown in Fig.67. For this stratification, wave motions are possible with $m \leq 7$, however the region of existence of the wave regime with $m = 4$ includes areas with larger wave numbers and the stable motion is only for the maximal wave number $m = 4$ for this stratification. A characteristic feature of this stability diagram for the wave motions and its difference from the corresponding diagram for the case of neutral stratification $\delta T = 0$, besides the filtration of the higher modes, is a considerable shift of the regime with $m = 4$ to the right which means that a range $2.25 < \ln k_0^{-2} < 2.82$ appears when there are only regimes with $m = 2$ and 3.

Note the main result of the influence of the stable vertical stratification: a shrinking of the regions of the axisymmetric flow instability and the filtration of higher modes was noted not once. Hignett et al (1981) choose as parameter determining the vertical stratification the value of $\mu = \frac{H^*}{h_u}$ (see section 3.10). When it was increasing, they observed a shift of the stability region to the right and downward, similarly to the results of Fig.66. Similar results could be obtained from a generalization of the Eady theory by inclusion of the friction in the Ekman layers which was done by Barcilon (1964). The main results of this work is given in Fig 50 for homogeneously heated rotating annuli. Instead of the Taylor number here the parameter proportional to $\epsilon = \frac{\Delta T_h}{\Delta T_v}$ was used and instead of the thermal Rossby number, a parameter proportional to ΔT_v was used (all the temperature changes are measured within the fluid outside the boundary layers). When the external vertical temperature difference ΔT_v is increasing (and therefore the value of Ro_T) and the value of ϵ is decreasing (and Ta), all critical curves shift to the right and downward in Fig.50.

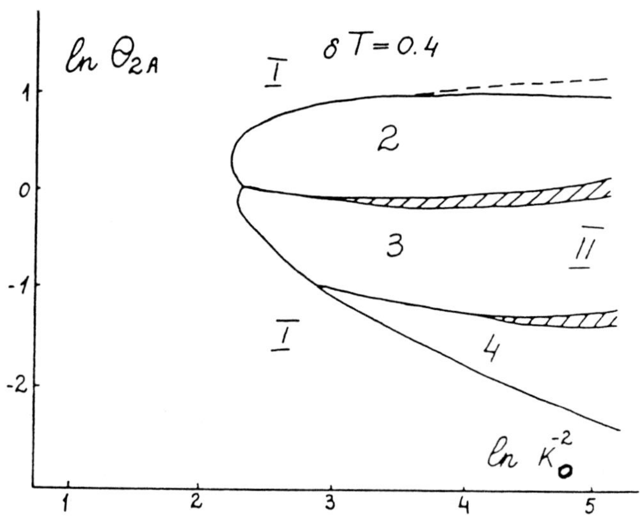

Figure 67: Stability diagram for $\delta T = 0.4$ and various values of the azimuthal wave number m shown by arabic numbers. I - zonal axisymmetric flows, II - regular wave flows. The hatched regions are regimes of adjacent wave numbers can exist simultaneously. (After Boubnov & Golitsyn, 1988, courtesy of Gordon and Beach Science Publishers).

However, for generalization of the results by Barcilon (1964) and Hignett et al (1981) on the case of using the external temperature vertical gradient difficulties arise in its turn at comparison of the model and experimental results. All the parameters introduced there for considering this influence (H^*, h_v by Hignett et al,1981 or ϵ by Barcilon,1964) are the internal parameters of the problem and depend in a complicated way on the external parameters of the system. Therefore, as in the case of our two-layer geostrophic model, one can may here only a qualitative comparison of the theory and experiments.

Experiments on the influence of the external stable stratification on the convective motions in rotating annuli were carried out by Boubnov and Golitsyn (1985,1988), Hathaway and Fowlis (1986), Miller and Fowlis (1986). The main difference in these experiments was in the values of the installation's geometric parameters: the value of $\frac{d}{b-a}$ of the first authors was 2.4 and 0.05-0.45 in the other works (with close values for a and b in all works). For the stable stratification, the results for deep and shallow vessels are similar. The determining parameter alongside with the Taylor and Rossby thermal numbers is the stability parameter

$$S_z = \frac{\alpha g d\, \delta T}{\Omega^2 (b-a)^2} \tag{4.38}$$

whose analog in the model is the value of δT. With the value of the parameter S_z increasing, the most pronounced changes are in the lower part of the regime diagram: the region of the wave motions shifts to the right and the axisymmetric motion becomes more stable. Also, the effect is observed (noted in the model) that the wave regimes with high wave numbers m shift stronger and that there are ranges of the Taylor number for which for the given geometry there is only the wave regime with $m = 3$ at all values of the thermal Rossby number observed in the experiments. Substantial differences for the deep and shallow vessel are observed for the case of unstable stratifications when $\delta T < 0$. But before we shall go to a consideration of this case in section 4.4 we shall first consider effects having a relation to this case of deviations from the hydrostatic and geostrophic approximations in the studies of the baroclinic instability.

4.3 Baroclinic instability without hydrostatics and geostrophy

Now we shall consider folowing Stone (1966,1970,1971), a generalization of the model by Eady (see section 3.1) by neglecting the hydrostatics and geostrophic approximation. The equations of motion in the Boussinesq approach for an inviscid incompressible fluid are

$$\frac{\partial u}{\partial x} + \frac{\partial v}{\partial y} + \frac{\partial w}{\partial z} = 0 \tag{4.39}$$

$$\frac{du}{dt} = 2\Omega v - \frac{1}{\rho}\frac{\partial P}{\partial x} \tag{4.40}$$

$$\frac{dw}{dt} = -2\Omega u - \frac{1}{\rho}\frac{\partial P}{\partial y} \tag{4.41}$$

$$\frac{dw}{dt} = -\frac{1}{\rho}\frac{\partial P}{\partial z} + \alpha g T \tag{4.42}$$

$$\frac{dT}{dt} = 0 \tag{4.43}$$

The fluid is assumed to be contained between two horizontal planes and the boundary condition then is:

$$w = 0 \quad \text{at} \quad z = 0, d \tag{4.44}$$

As the basic state we assume the zonal motion with a velocity u_0 and with a constant vertical shear homogeneous in the x-direction, with a constant stratification $\frac{\partial T_0}{\partial z}$ and a constant horizontal temperature gradient related to u_0 by the thermal wind equation (3.15). New non-dimensional variables are introduced with the scales: $\frac{u_0}{2\Omega}$ for x and y, d for z, $\frac{1}{2}\Omega^{-1}$ for t, u_0 for u and v, $2\Omega d$ for w, $d\frac{\partial T_0}{\partial z}$ for T, $\alpha g \rho d^2 \frac{\partial T_0}{\partial z}$ for P. Then the equation system (4.39)-(4.44) in non-dimensional form will be:

$$\frac{\partial u}{\partial x} + \frac{\partial v}{\partial y} + \frac{\partial w}{\partial z} = 0 \tag{4.45}$$

$$\frac{du}{dt} = v - Ri\,\frac{\partial P}{\partial x} \tag{4.46}$$

$$\frac{dv}{dt} = -u - Ri\,\frac{\partial P}{\partial y} \tag{4.47}$$

$$\gamma_0 \frac{dw}{dt} = -Ri\,\frac{\partial P}{\partial z} + Ri\,T \tag{4.48}$$

$$\frac{dT}{dz} = 0 \tag{4.49}$$

$$w = 0 \quad at \quad z = 0, 1 \tag{4.50}$$

where we have two new non-dimensional external parameters: the Richardson number

$$Ri = \frac{\alpha g d^2}{u_0^2}\frac{\partial T_0}{\partial z} \tag{4.51}$$

and

$$\gamma_0 = \frac{2\Omega d}{u_0} \tag{4.52}$$

which may be interpreted as the ratio of the geometric parameter $\frac{d}{L}$ to the Rossby number $\frac{u_0}{2\Omega L}$, where L is a horizontal scale. The basic stationary state is:

$$v = w = 0, \qquad u = z \tag{4.53}$$

$$T = z - \frac{y}{Ri} + const \tag{4.54}$$

$$P = \frac{z^2}{2} - \frac{yz}{Ri} + const \tag{4.55}$$

Now, we apply small perturbations to this basic state and linearize Eqs.(4.45)-(4.50). The coefficients of the linearized equations depend only on the vertical coordinate z. A solution of these equations is sought in the form of exp $[i\,(k_x x + k_y y + \lambda t)]$. The resulting equations for the amplitudes of the perturbations can be written as

$$ik_x u + ik_y v + \frac{dw}{dz} = 0 \tag{4.56}$$

$$i(\lambda + k_x z)\, u + w = v - ik_x\, Ri\, P \tag{4.57}$$

$$i(\lambda + k_x z)\, v = -u + ik_y\, Ri\, P \tag{4.58}$$

$$i\,\gamma_0^2(\lambda + k_x z)\, w = -Ri\,\frac{dP}{dt} + Ri\, T \tag{4.59}$$

$$i(\lambda + k_x z)\, T - \frac{v}{Ri} + w = 0 \tag{4.60}$$

$$w = 0 \quad \text{at} \quad z = 0, 1 \tag{4.61}$$

The system can be transformed to a single equation for the eigen values for the vertical velocity perturbation:

$$\begin{gathered} \left[1 - (\lambda + k_x z)\right]^2 \left(\tfrac{d^2 w}{dz^2}\right) - 2\left[\tfrac{k_x}{\lambda + k_x z} - ik_y\right]\left(\tfrac{dw}{dz}\right) - \\ \left[\left(k_x^2 + k_y^2\right) Ri - \left(k_x^2 + k_y^2\right)\gamma_0^2\,(\lambda + k_x z)^2 + \tfrac{2ik_x k_y}{\lambda + k_x z}\right] = 0 \end{gathered} \tag{4.62}$$

For the case $\gamma_0 = 0$, this equation can be transformed into the one obtained by Eady (1949) and also to Eq.(3.24). The analysis of this equation's solutions for $\gamma_0 = 0$ as performed by Stone (1966,1970) showed the existence of three main types of instability at $Ri > 0$:

(i) the "geostrophic" baroclinic instability which is really geostrophic at $Ri >> 1$; the main energy source for it is the potential energy of the unperturbed zonal flow and the maximal rate of growth of perturbations for this type of the instability is for $Ri > 0.95$;

(ii) symmetric instability (no azimuthal perturbations) exists for $Ri < 1$ and it takes the energy from the kinetic energy of the zonal flow, the maximal rate of growth is at $0.25 < Ri < 0.95$,

(iii) an instability similar to the Kelvin-Helmholtz one of two fluid layers with different densities and velocities (in the Eady model the change in density and velocity is continuous), maximal growth rate is for $Ri < 0.25$.

The value of $\gamma_0 = 0$ corresponds either to $\frac{d}{L} = 0$, or to infinite Rossby number and the deviation of γ_0 from zero can serve as a measure of deviation from zero rotation. The first interpretation is more preferable for the cases of geostrophyc and symmetric instabilities and the second one for the Kelvin-Helmholtz type instability.

4.3.1 "Geostrophic" instability

The maximal growth rate here exists for perturbations which do not depend on the lateral direction ($k_y = 0$), have a large scale in the zonal direction ($k_x << 1$) and at the same time have a rate such that $k_x \approx \lambda$. Then equation (4.62) reads

$$\left[1 - k_x^2(c' + z)^2\right]\left(\frac{d^2w}{dz^2}\right) - \frac{2}{\lambda + z}\left(\frac{dw}{dz}\right) - \left[k_x^2\, Ri - k_x^4\, \gamma_0^2\, (c' + z)^2\right] = 0 \qquad (4.63)$$

Then, the expansion is carried out on degrees of k_x^2:

$$w = w_0 + k_x^2\, w_1 + ..., \qquad c' = c_0 + k_x^2\, c_1 + ... \qquad (4.64)$$

Stone (1970) showed that the account for only two terms in the expansion (4.64) gave an error in the rate of the maximal perturbation growth no more than 7 per cent from the complete expansion; therefore we will keep only the first two terms.

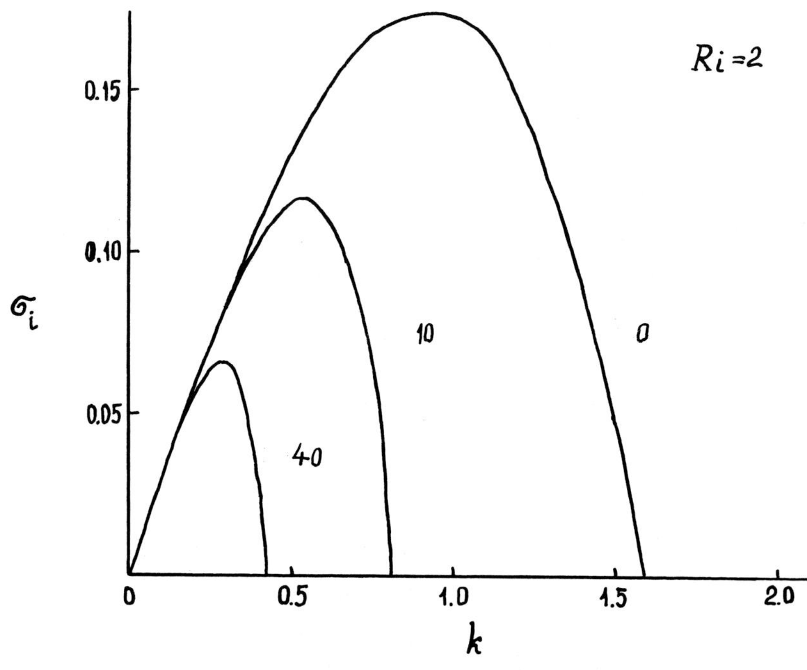

Figure 68: Growth rates against zonal wave number for "geostrophic" instability using hydrostatic ($\gamma_0 = 0$) and non-hydrostatic ($\gamma_0 = 10, 40$) conditions. (After Stone, 1970, courtesy of Cambridge University Press).

An approximate value for the rate of the perturbation growth can be obtained in this way:

$$\lambda_i = -Im\ \lambda \simeq 12^{-\frac{1}{2}} \left[k_x - 2k_x^3 \frac{1 + Ri + \frac{5\gamma_0 k_x^2}{42}}{15} \right] \tag{4.65}$$

Fig.68 presents the curves for the increments λ_i in dependence on k_x for three values of γ_0, i.e. 0, 10 and 40 at $Ri = 2$. The dependence on the Richardson numbers does not change qualitatively these curves. With the value of γ_0 increasing (the hydrostatics being less valid), the instable wave number and the growth rate are decreasing. The maximum growth is for the wave number determined by

$$k_{x,\max}^2 \simeq \frac{6.3}{\gamma_0^2} \left[\frac{2Ri - 2}{5} + \sqrt{\left[\frac{4\,(1 + Ri)^2}{25} + \frac{20\,\gamma_0^2}{63} \right]} \right] \tag{4.66}$$

The maximal deviation from the hydrostatics ($\gamma_0 \to \infty$) gives

$$\lim_{\gamma_0 \to \infty} k_{x,\max} = \left(\frac{63}{20} \right)^{\frac{1}{4}} \left(\frac{2}{\gamma_0} \right)^{\frac{1}{2}} \approx \frac{1.88}{\sqrt{\gamma_0}} \tag{4.67}$$

and the corresponding maximal growth rate is

$$\lim_{\gamma_0 \to \infty} \lambda_{i,\max} = \frac{1}{6} \left(\frac{63}{20} \right)^{\frac{1}{4}} \left(\frac{2}{3\gamma_0} \right)^{\frac{1}{2}} \approx \frac{0.54}{\sqrt{\gamma_0}} \tag{4.68}$$

4.3.2 Symmetric instability

Maximal growth rates for symmetric instabilities correspond to $k_x = 0$, $k_y >> 1$ and have the rates $\lambda_i \simeq 1$. The Eq.(4.62) at $k_x = 0$ has the form of

$$(1 - \lambda^2) \frac{d^2 w}{dz^2} + 2\,i\,k_y \frac{dw}{dz} - k_y^2 \,(Ri - \lambda^2 \gamma_0^2)\, w = 0$$

and the solution satisfying the boundary condition (4.60) is:

$$w = \exp \left[-\frac{i k_y z}{1 - \lambda^2} \right] \sin n\pi z \tag{4.69}$$

$$\lambda^2 = \frac{1 + \frac{1}{2}x^2(Ri + \gamma_0^2) \pm \sqrt{\frac{1}{4}x^4(Ri + \gamma_0^2) + x^2 + x^4\gamma_0^2(1 - Ri)}}{1 + x^2\gamma_0^2} \tag{4.70}$$

where $x = \frac{k_y}{n\pi}$, $n =$ 1,2,3,... . Only those eigen values in (4.71) with the minus sign before the square brackets correspond to the instability ($\lambda^2 < 0$). The growth rate is maximal at $x \to \infty$ and equal to

$$\lim_{x \to \infty} \lambda = \frac{Ri + \gamma_0^2 - \sqrt{(Ri + \gamma_0^2) + 4\gamma_0^2(1 - Ri)}}{2\gamma_0^2} \tag{4.71}$$

This means that the symmetric instability can exist only for $Ri < 1$ independently of the deviation from the hydrostatic balance. The maximal growth rate at the hydrostatic balance is equal to

$$\lim_{\substack{x \to \infty \\ \gamma_0 \to \infty}} \lambda^2 = 1 - Ri^{-1} \tag{4.72}$$

and at very large deviations from the balance

$$\lim_{\substack{x \to \infty \\ \gamma_0 \to \infty}} \lambda^2 = \frac{Ri - 1}{\gamma_0^2} \tag{4.73}$$

As for the case of the "geostrophic" instability, the growth rates of the symmetric instability decrease at increasing deviations from the hydrostatic balance. However, this influence starts to be felt considerably earlier ($\gamma_0 \simeq 1$, not only for $\gamma_0 >> 1$).

For finite values of x and k_y, the rates decrease falling monothonically to zero at

$$x^2 = \frac{1}{1 - Ri} \tag{4.74}$$

The general dependence of the growth rate on the value of x at $\gamma_0 \to \infty$ is:

$$\lim_{\gamma_0 \to \infty} \lambda^2 = \frac{1 - x^2(1 - Ri)}{\gamma_0^2 \, x^2} \tag{4.75}$$

4.3.3 Kelvin-Helmholtz instability

This type of instability is connected to sufficiently small-scale perturbations $k_x >> 1$ and has sufficiently large growth rates $\lambda \simeq k_x$. Let us introduce the geometrical parameter for perturbations in the form: $\varepsilon' = k_x \gamma_0$. In the limit of $k_x \to \infty$ and assuming that Ri, ε' and c' are all of order unity and $k_y \leq k_x$, then the Eq.(4.62) transforms into

$$\frac{d^2 w}{dz^2} + \left(1 + \frac{k_y^2}{k_x^2}\right)\left[\frac{Ri}{(z + c')^2} - (\varepsilon')^2\right] w = 0 \tag{4.76}$$

In this limit of large Rossby numbers, all effects related to rotation vanish. An analysis of the K-H (Kelvin-Helmholtz) instability for arbitrary flow profiles carried out by Stone (1970) showed that the K-H instability exists at $Ri < 0.25$. Here, the Richardson number should be interpreted as the local number, changing in (4.51) the value of $\frac{u_0}{d}$ by $\frac{\partial u}{\partial z}$. Quantitatively, the eigen numbers and growth rates depend on the profiles $\frac{\partial T z}{\partial z}$ and $\frac{\partial u}{\partial z}$ but qualitatively their behavior is the same. To the order of magnitude, the stablest are the perturbations for which very large azimuthal and zonal scales are comparable with vertical scales of the shear flow

$$k_y = 0, \qquad k_x \simeq \gamma_0^{-1} \tag{4.77}$$

Also their complex phase velocities are comparable with the rates of the basic flow $|c'| \simeq 1$. Therefore, to the order of magnitude the rate of the perturbations, growth is

$$\lambda_i \simeq k_x |c'| \simeq k_y^{-1} \tag{4.78}$$

until the Richardson number is not very close to 0.25.

According to Eqs.(4.62) and (4.77), at $\gamma_0 << 1$ and $k_x >> 1$ the rotation effects do not influence the K-H instability. Its rates of perturbation growth are considerably larger than those for the "geostrophic" and symmetric instabilities.

Fig.69 presents regions of existence of various instability types in coordinates (Ri, γ_0). The horizontal line $Ri = 1$ separates the region below which the symmetric instability can exist. The dashed curve corresponds to the parameter values for which the growth rates are equal for the "geostrophic" and symmetric instabilities.

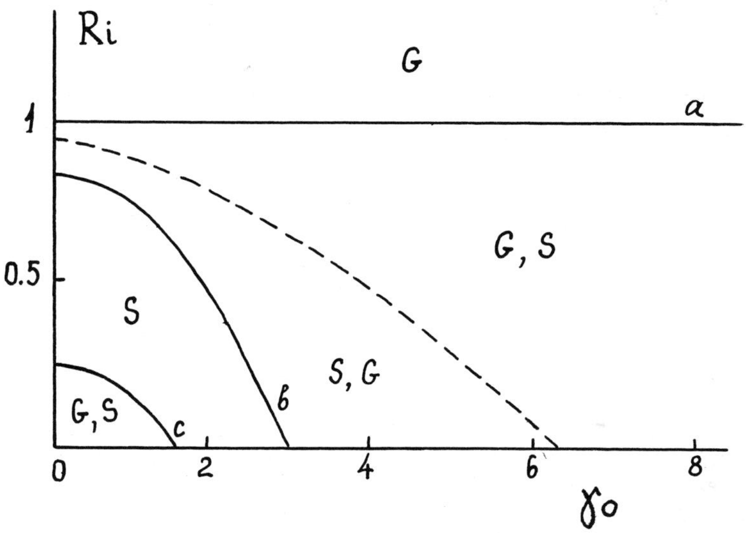

Figure 69: Instability regimes in the Ri, γ_0 plane. G stands for "geostrophic" baroclinic instability, S for symmetric instability, and K for Kelvin-Helmholtz instability. Along the dashed curve, the maximum growth rates for G and S are equal. (After Stone, 1971, courtesy of Cambridge University Press).

Let us pay attention to problems of the symmetric instability theory development. The consideration mentioned above by Stone (1966, 1970, 1971) was the first study which took into account three-dimensional perturbations. For the first time, this type of instability was considered for two-dimensional perturbations by Soldberg (1936) and generalized by Kuo (1956). The further development of the theory for the symmetric instabilities was carried out for the two-dimensional perturbations (McIntyre,1970) as well as three-dimensional perturbations (Busse & Chen,1981) with an account for dissipative terms. It was shown (see McIntyre,1970, Antar & Fowlis,1982) that the critical Richardson number for which the symmetric instability arises near $Ri \approx 1$ substantially depends on the Prandtl and Taylor numbers and has a minimum at $Pr \approx 1$ for a fixed value of Ta and is decreasing with Ta at fixed Pr. Busse & Chen (1981), for small values of k_x, showed that in the limit of infinitely large Taylor numbers the symmetric instability modes are inclined at some angle to the basic zonal flow chosen usually from the Eady problem. The sign of the slope for these modes in relation to the basic flow depends on the value of the Prandtl number: is it larger or smaller than unity. More realistic flow profiles were considered by Antar & Fowlis (1982,1984). In the second paper, it was found that for $Ta = 2.25 \cdot 10^7$ and $Pr \neq 1$ the "geostrophic" baroclinic instability prevails at $Ri > 0.825$ and the symmetric one at $Ri < 0.675$ in the sense of the growth rates. For large Taylor numbers ($Ta > 6.2 \cdot 10^8$) and $Pr = 1$, in the range $0.3 < Ri < 0.8$ the "geostrophic" instability always prevails. It was also found that the symmetric modes with the maximal growth rates are not completely symmetric but have some weak zonal structure.

4.4 Unstable vertical stratification

The symmetric baroclinic instability and the Kelvin-Helmholtz instability (see Fig.69) arise at positive but not too large values of the Richardson number, which corresponds to a weak but stably stratified external temperature stratification. One of the assumptions for an analysis of the baroclinic instability is the condition of constancy of the vertical temperature gradient within the fluid, i.e. the conservation of the external vertical temperature gradient. However, the presence of any horizontal temperature gradient leads to the origin of motion and such a motion causes an appearance in the fluid of the vertical gradient larger than the external gradient. It can be traced especially clearlyat the analysis of motions in an annulus in the absence of any external temperature gradient (see Chapter 3). Therefore, at a comparison of the theoretical results with experiments with regard to determination of the Richardson number, one should take the vertical gradient inside the fluid outside the boundary layers. The gradient will be small only at sufficiently large negative values of the external temperature gradient.

We shall continue the use of the two-layer geostrophic model which takes into account the vertical and horizontal inhomogeneous heating (see section 4.2) for the case of an unstable external stratification: $\delta T < 0$. We introduce an analog of the Richardson number:

$$Ri^* = \frac{3\sigma_0}{\tau^2} \tag{4.79}$$

where, as in section 3.7, the value of σ_0 determines the temperature difference between the layers and τ characterizes the shear between them. Taking into account Eqs. (3.58), (4.25) and (4.37) for an axisymmetric regime, or the Hadley regime, we have

$$Ri^*_H = \frac{3\sigma_0}{\tau^2} = \frac{3\sigma_0}{\theta_A^2} = P^* + \frac{\delta T}{\theta_A^2} \tag{4.80}$$

and for the wave regime, or the Rossby regime:

$$Ri_R^* = \frac{3\sigma_0}{\theta_A^2 + \theta_K^2 + \theta_L^2} = \frac{3\sigma_0\,\theta_A^2\,(Ri^* - 1)}{2(1 - V^*)} \tag{4.81}$$

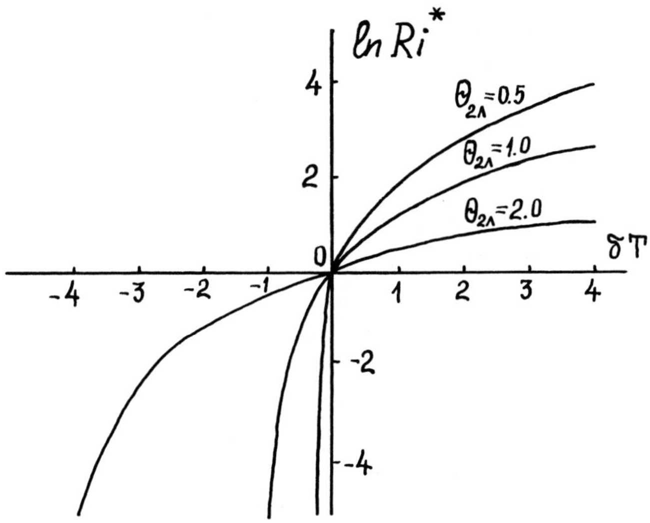

Figure 70: Dependence of the Richardson number Ri^* on the vertical temperature difference δT for the zonal flow for various values of the amplitude of the horizontal temperature difference θ_{2A} (shown on the curves). (After Boubnov & Golitsyn, 1988, courtesy of Gordon and Beach Science Publishers).

In the framework of this model, it is not possible to carry out any quantitative comparison of the value Ri^* with a real Richardson number. However, following the classification of section 4.3 one can suppose that some critical numbers Ri_1^* and Ri_2^* should exist such that for $Ri^* > Ri_2^*$, the "geostrophic" baroclinic instability should prevail, for $Ri_1^* < Ri^* < Ri_2^*$ the symmetric instability would be the main type and for $Ri_1^* > Ri^*$ the Kelvin-Helmholtz instability would be dominant.

Fig.70, in semilogarithmic coordinates, presents the dependence of Ri^* on the external vertical temperature gradient δT, i.e. the dependence of the model internal Richardson number on the model external Richardson number, for different values of the horizontal temperature difference θ_{2A}. This figure is an illustration of the fact that for motions with the internal Richardson number smaller than unity, one should use a negative value of the external Richardson number.

The diagram of existence of various regimes for the negative external vertical gradient $\delta T = -2$ is given in Fig.70. Qualitatively, the behavior of the fluid does not change strongly at large values of θ_{2A} and k^{-2} (right upper part of the diagram):

regions somewhat increase in which two wave regimes can exist simultaneously. The motions change principally at small values of the external parameters. The lower part of the critical curve for transition from an axisymmetric regime to the wave regime changes from a sloping line (at the stable stratification) to a horizontal line corresponding to the value $\sigma_0 = 0$ (for the unstable stratification). This condition means (see Eq.(4.25)) that $\theta_{2A} = \sqrt{(-\delta T)}$. With an increase of the wave number the transition curve from the axisymmetric regime to the wave regime shifts to the left and tends to infinity at $m \to \infty$.

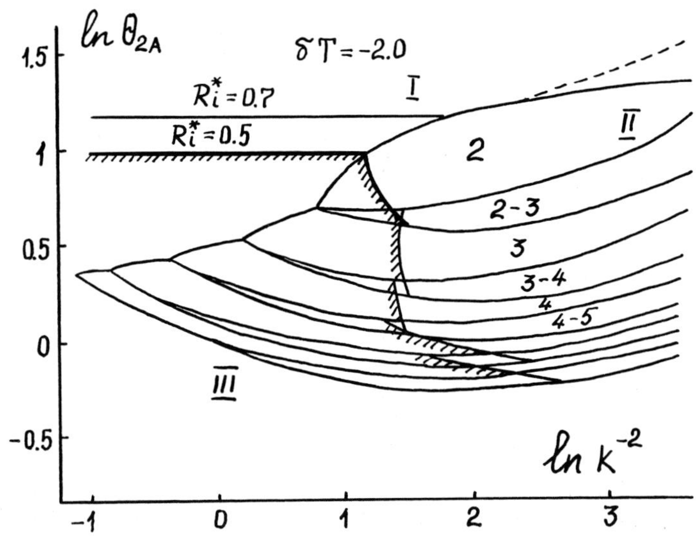

Figure 71: Stability diagram for $\delta T = -2$, showing isolines for the Richardson numbers $Ri^* = 0.5$ and 0.7: I - stable flow, II – eddy regime with various m, III - unstable regime due to Kelvin-Helmholtz instability. (After Boubnov & Golitsyn, 1988, courtesy of Gordon and Beach Science Publishers).

For unstable stratification there is no rotation rate for which at all values of the horizontal temperature difference θ_{2A} there is a stable axisymmetric circulation. And if the stable stratification filters out wave motions with large wave numbers m, the unstable stratification expands these regimes to the regions in which at the neutral external stratification there is the lower axisymmetric regime. For any wave number m^* at the negative values of δT, such a rotation rate k^{-1} can be found such that for all the rates smaller than k^{-1} the model will have the wave regimes with wave numbers $m > m^*$. But an existence of motions with large wave numbers at small values of the external parameters seems to be physically improbable and is caused by the impossibility to consider in the model the vertical structure in a more realistic and suitable fashion.

The introduction of the model Richardson number changes the regime diagram. In Fig.71 two curves are drawn corresponding to two different values of the Richardson number Ri^* calculated according to Eqs. (4.81) and (4.82). If we choose the critical Richardson number as $Ri_1^* = 0.5$ (somewhat arbitrary), then below the curve $Ri^* = 0.5$ the Kelvin-Helmholtz instability should develop. This instability in experiments should cause a mixing of the fluid and, as a consequence, the motion in an annulus should correspond to convective motions in a plane horizontal layer in the regime of geostrophic turbulence (see Chapter 2).

Experiments at unstable vertical stratification were carried out also in "deep" vessels: $\frac{d}{b-a} = 2.4$ (see Boubnov,1986, Boubnov & Golitsyn,1988) and in "shallow" vessels: $\frac{d}{b-a} = 0.05 - 0.45$ (see Hathaway & Fowlis,1986, Miller & Fowlis,1986, Stone et al,1969, Koschmider & Lewis,1986). These experiments have different aims. The goal of the experiments in a deep annulus was to explore the ranges of the external parameters at which there is a transition from the convection caused by baroclinic instability to the Rayleigh-Benard convection. For the corresponding values of the geometry parameters, the symmetric instability can not be studied. The baroclinic instability is observed in the experiment in a rather large parameter region changing only weakly at $S_z = -1.53$ as compared to the neutral stratification case $\delta T = 0$ in the upper right part of the regime diagram, i.e. the upper symmetric regime is observed as well as the regular wave regimes with various wave numbers $m = 2, 3$ and 4; see Fig. 72 where $G = \frac{a\Omega^2}{g}$. Principally the lower part of the diagram changes; there, instead of the lower axisymmetric and wave regimes the thermal convection is observed to be influenced by rotation. The arising flow pattern corresponds to the convective motion patterns in a plane rotating horizontal layer of the fluid with a homogeneously heated bottom in the regime of geostrophic turbulence. We also note in these experiments a development of a horizontal boundary between the axisymmetric regime and the regime of the thermal convection. This boundary is an additional confirmation of the possibilities of a few mode approximation model for the description of sufficiently complicated effects.

Similar results and the regime diagram were also obtained by Hathaway & Fowlis (1986) for a shallow vessel. The only qualitative difference is the existence there of baroclinic waves at large values of the Taylor number which evidently is related to the shallowness of the vessel because even the experiments at neutral stratification in the deep vessel show the presence of irregular geostrophic turbulence in this parameter region.

Note the experiments by Boubnov (1984) in which vertically and horizontally inhomogeneous heating was used for modelling the atmospheric circulation in the Southern Hemisphere. The experiments were carried out in an annulus with a heated outer cylinder and also a heated ring region at the vessel's bottom. The cooling was performed at the upper surface of the vessel. In dependence on the rotation rate and on the temperature at the bottom and at the outer cylinder, it was possible to have a two-cell circulation: the fluid was rising near the side walls and decreasing in the central part of the vessel gap, imitating the direction of the convection over the heated ring. Due to the Coriolis force action on these two cells, a shear flow was arising over the ring. As a result, it was possible by changing the system's rotation rate and heating to observe a transition from the thermal convection of the plane layer through the shear instability of the rotating fluid to the usual "geostrophic" baroclonic instability.

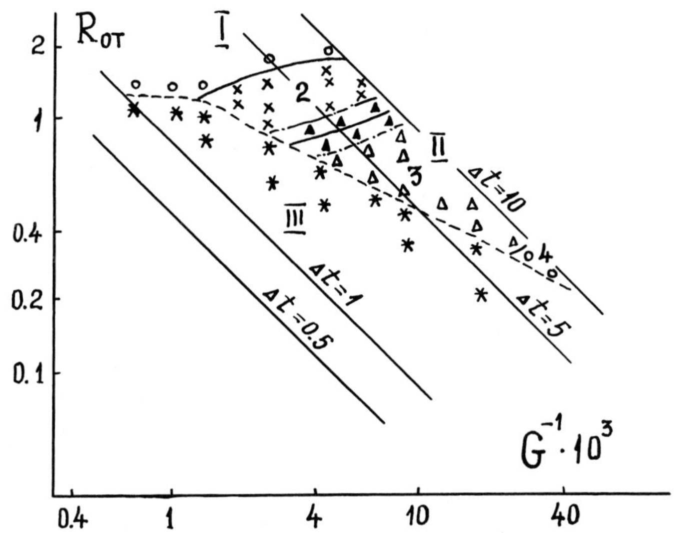

Figure 72: The experimental regime diagram for $S_z = -1.53$: I - stable flow, II - eddy regime with various m, III - the region of regular flows (stars); the dashed curve separates the irregular regimes from the regular ones. (After Boubnov & Golitsyn, 1988, courtesy of Gordon and Beach Science Publishers).

4.5 Diffusion circulation

Finishing the consideration of baroclinic instability in annuli with an inhomogeneous heating, we now shall consider flows in a fluid with a vertical density gradient developed not due to thermal causes but due purely to density reasons, e.g. by the use of solutions with different concentrations of soluble materials.

It is usually supposed that in a rotating fluid in which there is a gradient of temperature or of chemical concentration, it is impossible to have a solid body rotation (or a state of rest of the fluid relative to walls of a vessel rotating with a constant rate). This point of view is based on the circumstance that if there is a slope between surfaces of constant density and constant pressure, then, according to Eq.(1.13), the term determining buoyancy forces differs from zero and in the system there will be always a relative motion. If the surfaces, i.e. isopicnes and isobars, are parallel to each other then the solid rotation can exist in the system. We shall now consider possible variants of the mutual relative positions of isopicnes and isobars.

In inhomogeneously heated annuli discussed in Chapters 3 and 4 at a constant heating of the side walls, there is always a slope between isobars and isotherms (which in this case correspond to isopicnes). This slope determines the origin of an axisymmetric circulation, the stability or instability of which produces the main flow regimes presented in Fig.47. We remind the reader that partitioning of the axisymmetric regime on the upper and lower ones is caused by different mechanisms. The upper axisymmetric regime is a consequence of the Coriolis force action on the convective cell arising due to the ascending of the fluid at a warmer cylinder and its descending at a colder one. Here the dissipative factors, friction, and heat conduction play a minor role. The slope between isobars and isopicnes exists due to the arising sloping of the isotherms caused by the intensive convection; an additional curvature of the isotherms due to heat conduction can be neglected.

The lower axisymmetric regime is, vice versa, the direct consequence of the dissipative factors, namely of the heat conductivity. Therefore, the regime could be obtained in the theory only after introducing to the Eady model the dissipative factors, or, similarly, introduction of the exchange coefficient into two-layers models. In the case of the lower axisymmetric regime, the main cause of motions is the slope between the horizontal due to absence of an intensive convection isotherm and parabolic isopicnes. The sloping of the isopicnes (and isobars, which originally were parallel each other) is caused by centrifugal forces and in this regime it can not be neglected because it becomes the main cause of motions. The heat transport due to heat conduction leads to the fact that the horizontal density gradient is somewhat weaker and the isopicnes become not so curved as unchanging at constant rotation rate isobars. A slope is arising between these two surfaces and, consequently, an axisymmetric circulation arise relative to the vessel walls.

Precisely the same mechanism is causing the zonal circulation in a rotating fluid with a density stratification and it is called the Eddington-Sweet circulation. In laboratory experiments, the most often used is the density stratification of a fluid consisting of two layers with different densities, or a fluid with a linear vertical density gradient produced by a consequent filling of the vessel by fluids of different density. Usually, the density is varied by varying the concentrations of the salt or sugar solutions in water.

Suppose that in a vessel with depth d we have a linear density gradient $\frac{\Delta\rho}{d}$ where $\Delta\rho$ is the density difference over the whole depth. If the vessel is rotating with the constant rate Ω then in the vessel the isopicnes and isobars are parabolic surfaces due to the centrifugal force action. Due to the diffusion of salt in the horizontal direction (at one horizontal level there are fluid particles with different concentration due to parabolic isopicnes), the salt concentration is becoming more homogeneous in the horizontal and this is leading to the formation of a slope between surfaces of equal density (salinity) and equal pressure. As a result, the Eddington-Sweet, or diffusional circulation arise. For the thermally inhomogeneous fluid, the direction of the circulation is determined by the sign of the external temperature gradient: at the warmer outer cylinder and the colder internal one, the fluid rotation is always faster than the rotation of the vessel. For the case of the stable density stratification, the fluid is also rotating faster than the vessel.

Experimental studies by Buzina & Veronis (1971) and Griffiths & Linden (1985) have shown that such circulation does exist and its stability depends on the values of the external non-dimensional parameters. As parameters we take the Taylor number and an analog to the thermal Rossby number the internal Froude number:

$$Fr_i = \frac{4\Omega^2 R_0^2}{gd\frac{\Delta\rho}{\rho}} = \frac{4}{Ro_T} \tag{4.82}$$

where R_0 is a horizontal scale and the thermal Rossby number is determined as in Eq.(3.2) with $R_0 = b - a$ and $h = d$. Sometimes, another parameter is used which determines the relative role of the rotation and stratification: $\left(\frac{N}{f_c}\right)^2$ where N is the Väsäla-Brunt frequency and f_c is the Coriolis parameter. Then

$$\left(\frac{N}{f_c}\right)^2 = \frac{g\left(\frac{\Delta\rho}{\rho}\right)}{4\Omega^2 d} = \left(\frac{R_0}{d}\right)^2 \frac{1}{Fr_i} = \left(\frac{R_0}{2d}\right)^2 Ro_T \tag{4.83}$$

It was shown by Buzina & Veronis (1971) that the diffusion circulation can be baroclinically unstable and, instead of the axisymmetric circulation, baroclinic waves can arise. Griffits & Linden (1985) observed, beside the usual baroclinic waves, periodical oscillations of the diffusion surface for the layers of fluid with different densities having large periods from 2 to 4 hours and also the same type of oscillations after small changes of the rotation rate Ω (spin-up or spin-down).

Now we consider after Boubnov & Linden (1994) in some more details the effects which may arise at the instability of the diffusion circulation. Phenomena at spin up of the density stratified fluid from the state of rest up to the stationary rotation are sufficiently complicated. Therefore, to avoid such effects the linear stratification was produced in a vessel already rotating with a constant rate Ω and to do this the fluid with a linearly time decreasing density was introduced into the Ekman layer near the bottom of the vessel. The vessel was filled up during about two hours and after the filling ended, the diffusion circulation was established during a few minutes which, after dumping perturbations created during the filling-up, became axisymmetric after 20 to 30 minutes. Depending on the stratification N, which changed in the limits from 0.58 to 2.0 rad/sec, and on the angular rotation rate Ω, which was varying from 0.2 to 2.0 rad/sec, different regimes were realized in the fluid.

The regimes are presented in Fig.73 in the coordinates equivalent to the ones in Fig.47. In the left part of the diagram, there is a regime of a stable axisymmetric circulation which was not changing during of the experiment of 20 hours duration. If the angular rotation rate Ω increases, then the fluid velocity in this regime also increases and starting with some critical value Ω the circulation becomes unstable and after several hours of transition a wave regime arises in the fluid. At increasing the stratification parameter N, the fluid circulation velocity is also increasing, however, the stratification increase may end the circulation altogether and the fluid would be at rest relative to the vessel walls. We shall call such a regime the "solid body rotation". In the intermediate areas between these three main regimes some mixed irregular regimes may arise. Now, we shall consider these three regimes in some detail. In particular, we take the velocity point of view whose evolution seems to be the main factor which determines the establishment of one or another regime.

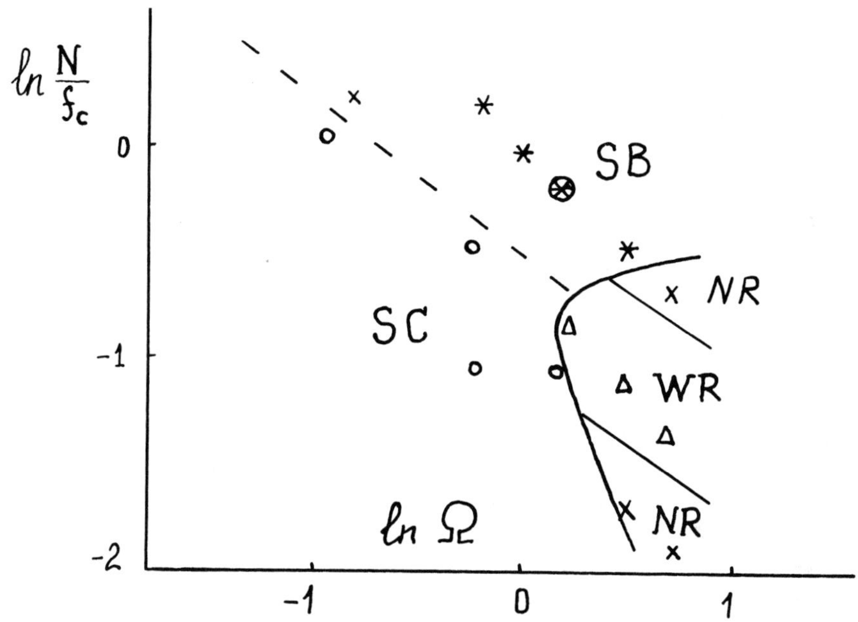

Figure 73: The regime diagram for diffusion circulation. Regimes: SC - symmetric circulation, SB - "solid body" rotation, WR - wave regimes and NR - nonregular regimes. (After Boubnov & Linden, 1994)

4.5.1 Axisymmetric regime

This regime is an analog of the lower symmetric regime for the fluid in a rotating annulus heated inhomogeneously at the radius. The velocity field there does not change with time and the maximum of the zonal velocity is about in the middle of the distance between the center of the vessel and its edges. Due to friction at the bottom, the maximal velocities are at the upper free surface of the fluid. A stationary pattern of the velocity field implies stationary mutual positions of the equal pressure and equal density surfaces, i.e. of isobars and isopicnes which were established sufficiently fast after the diffusion process started.

4.5.2 Wave regime

The wave regime takes place at large rotation rates and intermediate values of the density stratification. After the axisymmetric circulation arising in a certain time span, it becomes unstable and baroclinic waves and vortices appear which interact with each other for a time and then disappear leading again to the axisymmetric circulation. Such a process becomes periodical. The cause for such periodical transformations seems to be a periodical change of the horizontal density distribution which depends on the velocity shear in the zonal flow.

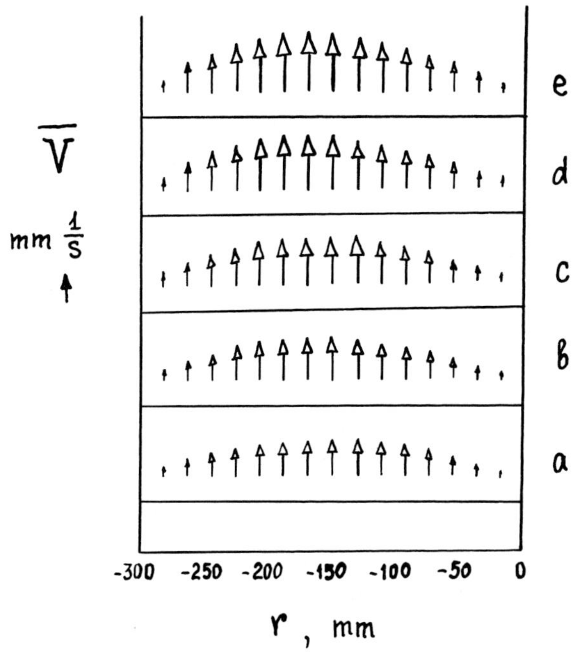

Figure 74: One period of the time-evolution of the radial velocity distribution for the wave regime (WR) of diffusion circulation. Time after the start of the symmetric circulation: a - 10 min, b - 20 min, c - 30 min, d - 40 min, e - 50 min. (After Boubnov & Linden,1994).

Now we consider in detail one of cycles of this periodic process which was first noted and studied, evidently, by Boubnov and Linden (1994). As an initial moment we take the time of the axisymmetric circulation formation after baroclinic waves and vortices cease to exist. Such a circulation has the zonal velocity maximum near the center of the vessel (see Fig.74a). Then the zonal velocity at points to the left of the first maximum at Fig.74b-e increases but the velocity near the center practically does not change. One may suppose that this effect is related to the density field transformation caused by diffusion. A density change wave seems to arise and the angle between isobars and isopicnes changes with time which leads to a gradual increase of the velocity field at the vessel periphery. The velocity maximum shifts gradually to the edges and at a moment the zonal circulation becomes unstable. Initial perturbations are in a form of baroclinic waves with the wave number $m = 2$ or 3, and its external appearance is as for the corresponding waves in a inhomogeneously heated rotating annulus. However, the waves also become unstable: they are transforming into a system of a small number (from one to four) of vortices of different intensity interacting with each other. The final results of such interactions and transformations, which are taking place during 30-50 minutes, is the formation of a symmetric vortex with the velocity maximum near the vessel center (see Fig.74a); this means that the motion transformation is such that a part of the kinetic energy goes to changes of the fluid potential energy and into heat.

The process of the zonal flow instability may acquire various forms. The pattern of the transformation depends on the position at the vessel height and its intensity is maximal at the fluid free surface were the fluid velocity is also maximal. E.g., at the surface in some experiments the axisymmetric circulation was breaking down into two or more vortices which were interacting during several minutes while at half the height only a small curvature of the zonal circulation was observed.

4.5.3 "Solid body rotation"

The behavior of the fluid at large values of the vertical stratification differs principally from the case of an inhomogeneously heated annulus. At initial stages, the process evolves as in the case of the wave regime. After filling up the vessel in a short time, the zonal circulation is established, the velocity maximum where is moving from the center to the edge of the vessel and the velocity is increasing. Near the edge, the strong stratification does not allow a development of a baroclinic instability and curvature of the isopicnes is decreasing at the approach to the edge, i.e. the slope between the isopicnes and isobars instead of being increasing as it was before starts to decrease from the edge. The situation is presented in Fig.75 where as the initial moment the time is chosen of the approach of the maximum to the edge which is equivalent to the last moment in Fig.74e. Due to this, the velocity starts to decrease near the edge and this decrease propagates to the center of the vessel. In some places, such a decrease of the slope leads to the formation of zones where the slope becomes negative , i.e. the horizontal changes of the density will have a maximum within the fluid. This leads to the development in some time of axisymmetric flows in opposite directions near the edges and near the center. After a while, all flows cease to exist and the solid body rotation arises when, probably, the angle between isobars and isopicnes changes sign with a period less than a time of diffusion circulation is developed. Such a transformation for the vessel with a diameter of 35 cm and a fluid depth of 20 cm took 1.5-2 hours after the end of the filling up the vessel and the solid body rotation was observed during more than 24 hours after which the experiment was terminated.

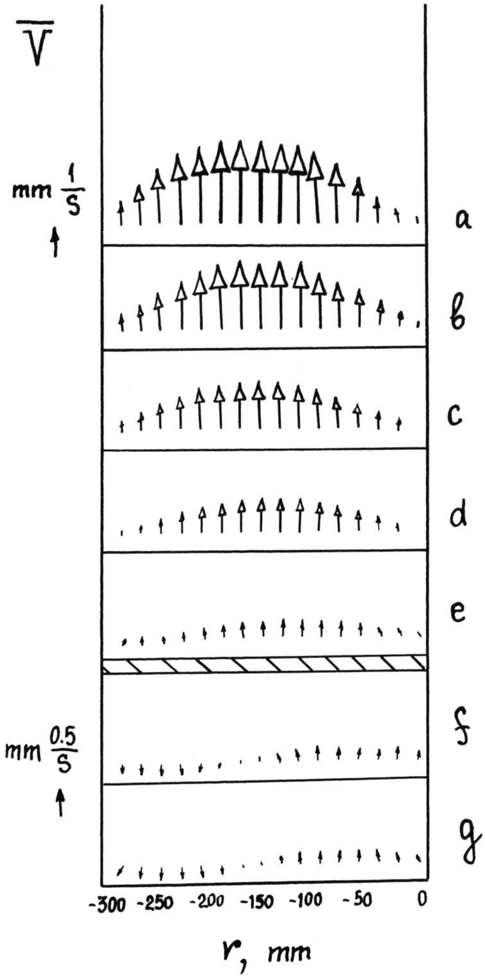

Figure 75: Time evolution of the radial velocity distribution for regime "solid body" rotation (SB). t : a - 0, b - 10 min, c - 20 min, d - 30 min, d - 40 min, e - 50 min, f - 55 min, g - 65 min, where t is the time from 105 min after the "stratification", when the situation of Fig.86e is established. Velocity scales change twice between e and f. (After Boubnov & Linden, 1994).

Irregular regimes shown in Fig.73 at the boundaries between the wave and other regimes are obviously a consequence of interactions between the regimes. For instance, in the upper part of the wave regime a point is marked where after damping the velocity fell by an order of magnitude; however, a weak curved circulation was existing. Similarly in the lower part of the wave regime at a weak stratification there was a curved zonal circulation in which the wave patterns were very weak.

Beside these regimes, in one experiment at the non-dimensional parameters corresponding to the solid body rotation (during the repeated experiments only such a regime was observed) in about 20 hours a formation of a strong zonal circulation was observed with a layered vertical structure. This may be an evidence of a possibility of such density field transformations when local homogeneous layers form parallel to isobars and the circulation driving force is the diffusion at the boundaries between the layers.

Estimates of the maximal velocities for the zonal axisymmetric circulation have shown that these velocities are proportional to the value of the Väsäla-Brunt frequency N and they increase at about the square root of the rotation rate Ω. Such a dependence is shown in Fig.76. The maximal circulation velocity observed in the experiments was up to 1 $cm\,s^{-1}$ and the maximal rate of the fluid rotation relative to the vessel walls was more than 3 per cent of the system's general rotation rate Ω.

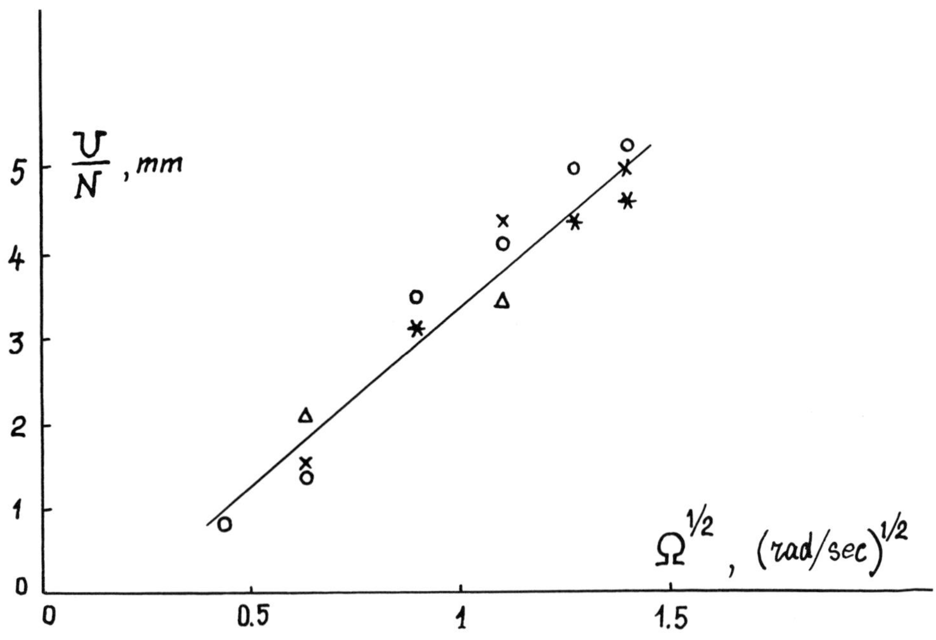

Figure 76: Dependence of maximal velocity (in N^{-1} units) on $\Omega^{\frac{1}{2}}$ for the different regimes of diffusion circulations: N ; $\times - 2.0$, $\bigcirc - 1.04$, $\triangle - 0.87$, $* - 0.58\ s^{-1}$. (After Boubnov & Linden,1994).

In conclusion, we note analogies and differences of the experiments on baroclinic instability at the pure density and thermal stratifications. The physical mechanisms leading to the formation of the axisymmetric circulation are the same. There is a gross similarity between the regime diagrams. The main difference lies in the appearance of the new regime of solid body rotation instead of the upper symmetric regime which is caused by the damping of vertical motions and by the impossibility of formation of a circulation cell in the stratified fluid in this diagram region. The second principle difference is caused by the staging of the experiments. In homogeneous rotating annuli (Chapters 3 and 4), the external temperature is fixed and the internal horizontal structure of the density field is determined substantially by this external thermal forcing while the horizontal structure of the density field in the stratified fluids experiences marked periodical oscillations.

Chapter 5
CONVECTION FROM LOCAL SOURCES

In the three previous Chapters 2-4 we have considered the two limiting cases of heating in the horizontal and in the vertical and their different combinations. These are the simplest cases and in a sufficiently wide range of the system external parameters there is a possibility to study them both from theoretical and experimental points of view. Geophysical applications originate from studies of convection in rotating spheres and spherical shells. The studies of regular motions from theoretical points of view differ little from considerations of a homogeneously heated plane horizontal layer (Chapter 2). However, experimental studies are much more complicated and the progress reached recently is connected with a complex and expensive equipment; e.g., to study the convection in a spherical shell with a body force imitating the Earth central gravity field the experiments have been conducted on the Space Shuttle by Hart et al (1986).

An opposite situation exists for studies of the convection from local sources in rotating fluids. Experimental equipment and measurement methods do not differ practically from experiments in a plane horizontal layer. Usually, a part of the heated surface is thermally insulated for the thermal convection or a local source of the density convection is build in. But any theoretical considerations of the convection from local sources is greatly complicated due to additional inhomogeneous boundary conditions at the source edge and the impossibility to carry out a simple expansion on horizontal eigen functions at the analysis of perturbations.

At considering convective motions in a plane horizontal fluid layer, it was assumed that the governing length scale is the layer depth h, which should be substantially less than a characteristic horizontal scale L ($L >> h$, or $\delta_1 = \frac{L}{h} >> 1$). In the theory this is an usual approximation for an infinite plane layer. In experiments, this condition is not always fulfilled but there is a range of parameters in which changes of δ_1 do not influence some results which can be compared with theory. This situation was considered in Chapter 2, e.g. at the investigation of the convection onset.

At the inhomogeneous heating in the horizontal, the value of δ_1 plays an important role in considering the stability for small values of the Richardson number (see section 4.3), where the parameter $\gamma_0 = (\delta_1 \cdot Ro)^{-1}$ is one of the important parameters and determines the regions of development of the symmetric instability and of the Kelvin-Helmholtz instability.

In the case of localized sources of buoyancy, there is an additional scale of length: the size of the source D which leads to the appearance of the new non-dimensional parameter $\delta = \frac{D}{h}$. Under localized sources we will assume a buoyancy source with size D such that it is substantially smaller than a characteristic horizontal size of the system L, i.e. $D << L$. We consider first convective motions from a localized source in a non-rotating fluid and then the case of rotation at a range of external determining parameters.

5.1 Local buoyancy source in a non-rotating fluid

The main determining non-dimensional parameters here for the localized convection in a plane horizontal infinite layer, or $L >> h$, are the Rayleigh number Ra, Prandtl number Pr and the spatial parameters $\delta = \frac{D}{h}$. The choice of the proper length scale in the determination of the Rayleigh number, h or D, depends on the value of δ. We define two related values of the Rayleigh number:

$$Ra_D = \frac{\alpha g D^3 \triangle T}{k\nu} = \delta^3 \, Ra_h \tag{5.1}$$

If $D << h$, or $\delta << 1$, the main determining parameter becomes Ra_D and the problem is that of the localized source in the upper infinite semispace. Such a problem can be considered for two aspects: (i) origin and development of convection near the heated surface at distances $z < O(D) << h$, and (ii) the development of motions far away from the buoyancy source $O(D) << z << h$.

In the studies of the flow field near the source an important role is played by the geometry of the system. Fowlis & Hide (1965) considered various cases of placing the heat source (elevated above the plane, deepened within the plane etc.), which revealed a good correspondence between experimental and numerical results. In the general case, the main characteristics in which differs the local heat source from an infinite heated plane is the presence of regions in the fluid unperturbed by convection, the motion in which is determined by the fluid behavior at the boundaries of the buoyancy flux, usually having a form of an upward expanding jet. Near the source, motions arise directed horizontally to the source center caused by the fluid outflow on the heated part of the plane. It is only natural that for the laminar regime for such a motion the geometry of the system is important.

For developed convection at large Rayleigh numbers, the lateral motion changes the flow pattern strongly near the edge of the heated source and even noticeably above the center. Experimental studies at $Ra_D > 4 \cdot 10^7$ for plane heated bodies of various sizes and shapes (circles, squares, triangles; see Husar & Sparrow,1968, Sparrow & Husar,1969, Fujii & Umara,1972, Al-Arabi & El-Ready,1976, Al-Arabi & Sakr,1976) showed that in the center over the source turbulent convection develops, not differing from the convection in the plane layer while near the plate edges the motion is laminar. The distance occupied by the laminar motions depends only weakly on the shape of the heated source, therefore regions in which turbulent convection develops are repeating the shape of the source (circles, squares, triangles) but are decreasing with height. In the region of transition from laminar to turbulent motions rather regular vortex structures are observed in which the horizontal axis are directed perpendicular to the edges of the heated parts. In a large part of these works the main attention has been paid to investigation of the heat transfer from heated surfaces to the surrounding fluid, to determination of the exponent β in the dependence of the Nusselt number on the Rayleigh number $Nu \propto Ra_D^\beta$. The value of β changes depending on the regime from $\beta = \frac{1}{5}$ for a laminar regime (or area) to $\frac{1}{4}$ in the transition regime, to $\frac{1}{3}$ in the turbulent regime (or area occupied by turbulence) of convection.

Studies of the flow far away from localized sources is a sufficiently well developed area of investigation of buoyant jets and thermals. The main process here is the entrainment of the surrounding fluid into jets and thermals and changes of their scales with height. The parameter range here is for $\delta << 1$ and layers of large depth. Sufficiently complete reviews on this problem were published by Turner (1973,1986), Gebhart (1973), Bill & Gebhard (1975).

Convection in a closed cavity with sizes comparable to the size of a heated surface was studied numerically and experimentally by Torrance (1979), Torrance & Rockett (1969). The main attention was paid to studies of convective cells induced by a jet rising from the buoyancy source. Such motions were studied in homogeneous as well as in stratified fluids.

For configurations with $\delta_1 << \delta << 1$, or for the length scale $L >> D >> 1$, one should choose the layer height h for determining the Rayleigh number. In this case, the problem becomes practically the problem of convection in a plane infinite horizontal layer (see section 2.6) with some additional effects arising on the edges of the heated plate, which at a large area of the plate, influence only weakly the convection near the center of vessel.

Most complicated is the case of intermediate values of $\delta \approx 1$, or $D \approx h$, when both scales are of importance. Boubnov & Heijst (1994) carried out experiments with and without rotation in a wide range of values of the parameter δ. For the fluid at rest and the case $L >> h, D$, the regime diagram is constructed summing up the above proposed classification in the coordinates Ra_D and δ with an additional axis $Ra_h = Ra_D \cdot h^3$. The diagram is presented in Fig.77.

In section 2.6 the turbulent convection has been considered in the system at rest and it was shown that there is a characteristic frequency τ_0 given by Eq.(2.56) and a corresponding spatial scale λ_0 (see Eq.(2.57)). This frequency is caused by a periodic break-up of the surface boundary layer. The first experimental finding of such a frequency was noted by Sparrow et al (1970) in whose study the temperature oscillations were purely periodical with frequency τ_0 and thermals were formed as mushroom shaped rising motions. These results were interpreted as a form of convection from an infinite heated plane. In further works, described in section 2.6 at the studies of turbulent convection, in an infinite layer irregular temperature fluctuations have been observed having a maximum of their spectrum at this frequency. The thermals there changed from mushroom-like structures and evolved with time into a system of irregularly rising jets.

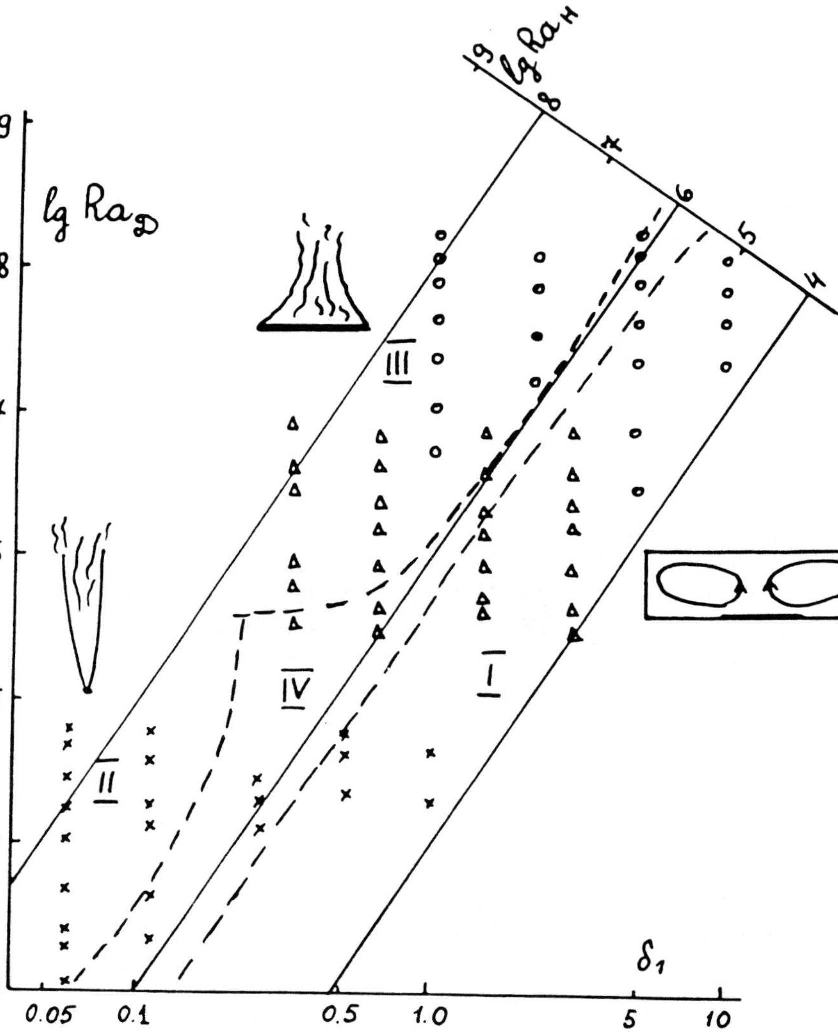

Figure 77: Regime diagram of the observed convective flow patterns on the parameters δ, Ra_D and Ra_H. The symbols refer to different plate diameters D: $\times - 1\,cm, \triangle - 6\,cm, \bigcirc - 20\,cm$. The various regimes denote: I - laminar toroidal cell, II – thermal plume, III - turbulent convection, IV- transition regimes. (After Boubnov & Heijst, 1994).

The contradiction between the results of Sparrow et al (1970) and subsequent works can easily be understood because in their work the authors had a disk with a diameter $D = 8.7cm$ mounted at the height $8cm$ from the bottom of their basin and this should be considered as a local source of buoyancy.

Boubnov & Heijst (1994) measured in detail the turbulent convection temperature field. Fig.78 presents a flow pattern in a turbulent regime from a local source which was a disk of 20 cm in diameter. Besides the mechanism described in section 2.6 of periodical break-up of the thermal boundary layer and formation of thermals, one could clearly see the fluid motion from the edge of the heated disk which leads to a vertical shearing of those thermals which form at some distance from the edge toward the center. The thermals which could form near the edge are transforming (due to the action of the suction flow towards the disk) into wave motions near the edge. These waves develop independently on the distance from the edge and could be stable or overturning. Similar wave motions were observed in a region of transition from the laminar convection to the turbulent one near the edge of an inclined heated plate (Sparrow & Husar,1969, Lloyd & Sparrow,1970, Fujii & Umara,1972; Black & Norris,1975).

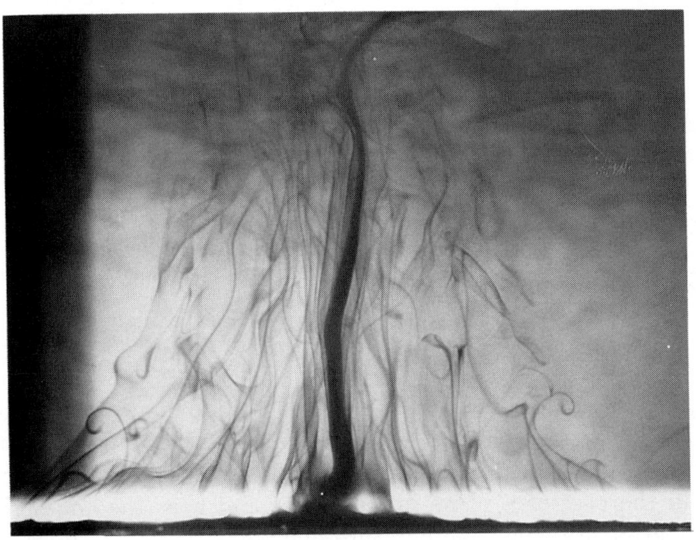

Figure 78: Typical convective motion in a turbulent regime. $D = 20\,cm$, $H = 17.6\,cm$, $\triangle T = 7.6\,K$. (After Boubnov & Heijst, 1994).

Measurements of the temperature at various levels over the heated surface showed that the period of the temperature pulsations is the same for the wave motions as for thermal formation. These oscillations are especially clearly observed at the boundary of the rising flow whose horizontal cross-section decreases with the height under the convergent fluid motion from the disk edges to its center. Closer to the central part of the disk, these oscillations become less regular and over the center of the heated disk the temperature fluctuations are equivalent to the fluctuations over an infinite heated plate as described in sections 2.6.

The most probable frequency of the temperature fluctuations, corresponding to the temperature spectrum maximum, was expressed through the Rayleigh flux number while more often the ordinary Rayleigh number was used. Using the relationship between the two numbers $Ra_f = Ra \cdot Nu$ and taking into account that for the turbulent regime of convection in the non-rotating fluid there is a simple dependence of the Nusselt number on the Rayleigh number ($Nu = C'_{Nu} \cdot Ra_f^{\frac{1}{3}}$), one can rewrite Eq.(2.54) as

$$\tau_0 = C'_\tau \cdot \frac{h^2}{\nu} \cdot Ra^{-\frac{2}{3}} \tag{5.2}$$

where $C'_\tau = \frac{C_\tau}{\sqrt{C'_{Nu}}}$.

Experimental tests of the dependence of the Nusselt number on the Rayleigh number for the turbulent regime carried out in various experimental studies reviewed by Al-Arabi & El-Ready (1976) give the value of the coefficient C'_{Nu} in the range 0.14-0.17.

The treatment of experimental results by Boubnov & van Heijst (1994) has shown that the dependence (5.2) is fulfilled well in the range of the Rayleigh numbers from 10^6 to $2 \cdot 10^{12}$ and the value of $C'_\tau = 36.1 \pm 2.2$. Assuming $C'_{Nu} = 0.155 \pm 0.015$, we obtained the value $C_\tau = 14.2 \pm 2.3$ for the regular oscillations over a local surface. This is close to the value $C_\tau = 12.7 \pm 1.5$ in section 2.6 for the most probable frequency of oscillations in turbulent convection over an infinite plane.

5.2 Geostrophic balance and local sources

The general consideration of convection from a local source in a rotating fluid should take into account all effects considered before plus all possible combinations of them. First, the problem of the convection onset. In the center of the source the convection should arise as in the case of a plane infinite layer and near the edges the motion should appear at arbitrary small temperature differences between the disk and the surrounding fluid. The problem of interactions between these motions is a very complicated one.

Similarly difficult is the problem of turbulent convection. In section 5.1 it was shown that for the non-rotating fluid the local source of buoyancy can cause a wide range of hydrodynamic phenomena: transition from laminar flows to turbulent ones, instability of jets in the fluid, convective instabilities. The influence of the system rotation will be different for various types of motions and scales and only recently a few works have appeared in which there were attempts to consider some of these effects (Boubnov & Heijst,1994, Maxworthy & Narimosa,1991, Maxworthy,1992).

The main non-dimensional parameter determining the influence of the rotation is, as usual, the Rossby number $Ro = \frac{U}{2\Omega L}$. Among the parameters determining the Rossby number for a local source only the rotation rate Ω is clearly determined. The criterium of the rotation influence is the condition $Ro < 1$, or, for the constant Ω, $\frac{U}{L} < \left(\frac{U}{L}\right)_{cr}$. For a local buoyancy source in a fluid at rest in deep fluid, $D << h$, the cross-section diameter of the rising fluid jet decreases first, but not more than 50%, and then starts to increase due to the turbulent entrainment of the surrounding fluid. The characteristic flow velocities are decreasing due to dissipative processes. As a result, a distance from the source always exists at which due to expansion of the rising flow we shall observe $\frac{U}{L} < \left(\frac{U}{L}\right)_{cr}$ and the influence of rotation will be substantial. The turbulent convection with rotation effects will arise and is equivalent to the one considered in Chapter 2.

If the layer is shallow, i.e. $L >> D >> h$, then a spread is essential of the rising fluid near its upper surface. Such a spread will be up to the side walls or, beginning from a scale L, the rotation will become important and the baroclinic instability will evolve as described in Chapters 3 and 4.

If the characteristic horizontal and vertical scales of the system are small then a situation would be possible where the influence of rotation could be insignificant as in cases of slow rotation or weak heating. For the last case, estimates are possible of the rotation influence similar to the one in section 4.1; however, one can not use the expression for the width of the thermal boundary layer proportional to $Ra^{\frac{1}{5}}$.

Now we consider some results of a rather small number of experiments with local buoyancy sources in the geostrophic approximation, i.e. at $Ro << 1$. Such experiments were performed mainly for the turbulent regime of convection. The buoyancy flux was produced both by a local heated source at the vessel bottom (Boubnov & Heijst, 1994) and by the source of the salted (and more dense) fluid permanently introduced at a small rate on the surface of the fluid (Maxworthy & Narimosa, 1991, Maxworthy,1992). The different sources of buoyancy have much in common but also important differences. The differences are that in the experiments with the more dense fluid source additional effects arise related to the diffusion circulation (see section 4.5) which in some critical situations can be determining not only by producing an additional general circulation in the vessel but could substantially influence the motion stability and the formation of various vortex structures and motions. Beside this, the permanent injection of the denser fluid introduces into the fluid volume an additional source of mass and momentum and such a system differs from the dynamically closed system in the case of thermal convection experiments. For some experimental conditions, an additional spread of the fluid at the bottom becomes substantial leading to new effects. In spite of these differences, the main results from both sources of buoyancy in the turbulent convection regimes are similar, and this tells us once more that these experiments are modelling the influence of rotation and buoyancy on the turbulent mixing in fluids irrespective on the sources of the mixing.

One of the expected results of the rotation's influence at small Rossby numbers is the principal change of convection flow patterns from those observed in the system at rest. The lateral flows from the edges of the heated plate contradict the conditions of the Taylor-Proudman theorem and instead of flow convergent and centrally rising and narrowing with height over a limited heated disk (Boubnov & Heijst, 1994), convective flows are forming similar to convection in a plane infinite layer, i.e. the vortex structure described in Chapter 2.

As an example of the density convection, we shall now consider the process of motion evolution with time after Maxworthy (1992). Note that as for many convective processes, the effects arising during the transition to the stationary convection and at stationary regimes are not always identical. An example of such a difference is the convection in a plane layer with no rotation when in the beginning the convection is in the form of sufficiently regular rising mushroom-like thermals which with time are transforming into convective jets. Therefore, a number of the results presented below are similar to analogous results for the stationary convection but some are transient: e.g. the depth z_{cr} of the transition from two-dimensional motions to fully three-dimensional ones may decrease due to entrainment of the axisymmetric (two-dimensional) motions into the mixing layer.

The buoyancy flux for the density convection is $B_0 = qg\frac{\triangle\rho}{\rho}$, where q is the density flux for a unit area, $\triangle\rho$ is the density excess of a heavier fluid over the working fluid. For comparison with the thermal convection, the value of B_0 has to be equivalent to the rate of the kinetic energy generation (dissipation) $\varepsilon = \frac{\alpha g f}{\rho c_p}$ for the case of $Nu >> 1$, i.e. for the well developed convection when the dissipative factors are of minor importance (see Chapter 2). More details on the analogy between density and thermal convection can one find in Golitsyn (1979,1980a).

If the depth of the layer is larger (e.g. $H > z_{cr}$) then the rotation starts its action on the turbulent fluctuations and the following scenario of the convection process evolution is possible. In the very beginning, for times $t < f_C^{-1}$ with f_C the Coriolis parameter, the turbulence generated is three-dimensional and does not depend on the rotation. The determining parameters are the flux B_0 and the vertical scale z from which a scale of velocity can be formed (i.e. $U \propto (B_0\, z)^{\frac{1}{3}}$) and an integral length scale $l \sim z$. The turbulence is entraining into the fluid and at the front of the entrainment area with vertical coordinate z_f one may determine the Rossby number

$$Ro_f = \frac{U}{f_C\, z_f} \propto \frac{(B_0\, z_f)^{\frac{1}{3}}}{f_C\, z_f} = \frac{B_0^{\frac{1}{3}}}{f_C\, z_f^{\frac{2}{3}}} \tag{5.3}$$

When this number will reach the order unity, one may expect an essential influence of the rotation on the turbulent fluctuations. This will be at the depth z_{cr} and after a time t_{cr}:

$$z_{cr} = C_z \frac{B_0^{\frac{1}{2}}}{f_C^{\frac{3}{2}}}, \qquad t_{cr} \propto \frac{z_{cr}}{(B_0\, z_{cr})^{\frac{1}{3}}} = \frac{l}{f_C} \tag{5.4}$$

Experiments by Fernando et al (1991) and Maxworthy & Narimosa (1991) gave a good agreement of the results for both thermal and density convections at checking the first equation (5.4) with the coefficient $C_z = 12.7 \pm 1.5$. The value of the critical Rossby number is found to be $Ro_f = 0.37$. For this regime the reduced gravity is

$$g' = g\frac{\Delta\rho}{\rho} \propto B_0^{\frac{2}{3}} \propto (B_0\, f_C)^{\frac{1}{2}}, \quad U_v = w \propto (B_0\, z_{cr})^{\frac{1}{3}} \propto \left(\frac{B_0}{f_C}\right)^{\frac{1}{2}} \tag{5.5}$$

which coincides well with a characteristic velocity for the geostrophic convection (2.60) first introduced by Golitsyn (1980) and the numerical coefficient in the formula (5.5) for the velocity for both types of convection agree well with each other. The vortex diameter can be written as $D_v = C_D B_0^{\frac{1}{2}} f_C^{-\frac{3}{2}}$. Experiments by Maxworthy & Narimosa (1991) confirm this dependence with $C_D = 15.0 \pm 1.5$. The vortices in these experiments reach the bottom of their vessel during the time of order $h \cdot B_0^{-\frac{1}{2}} f_C^{\frac{1}{2}}$. The further stage is the development of the baroclinic instability with the scale D_b of the order of the Rossby deformation radius (see e.g. Griffiths & Linden,1981):

$$D_b \propto \left[h\left(\frac{B_0}{f_C^3}\right)^{\frac{1}{2}}\right]^{\frac{1}{2}} = B_0^{\frac{1}{4}}\, h^{\frac{1}{2}}\, f_C^{-\frac{3}{4}} \tag{5.6}$$

or

$$\frac{D_b}{h} = C_{Db}\, (Ro^*)^{\frac{1}{2}}, \qquad Ro^* = \left(\frac{B_0}{f_C^3\, H^2}\right)^{\frac{1}{2}} \tag{5.7}$$

where the Rossby number is determined by the geostrophic convection scale and h. According to Maxworthy & Narimosa (1991), the value $C_{Db} = 5.2 \pm 1.0$.

According to results of experiments by Boubnov & Heijst (1994), Maxworthy and Narimosa (1991), Fernando et al (1991) the regime of the turbulent convection near the center of the local source is similar to motions in a plane infinite layer which is confirmed not only by the similar dependencies but also by coincidence of the numerical coefficients in them. The values of the Rossby number determined by the source diameter at which this confidence is observed is $Ro \leq 0.2$. At larger values of the Rossby number over the heated disk, the interaction of two mechanisms becomes essential: of the system rotation and of the fluid convergence from the disk edges, i.e. beside the balance of the pressure and Coriolis forces, inertial forces become important and these effects are considered below.

5.3 Intense solitary vortices

If there is no geostrophic balance in the fluid, motion from the edges of the heated plate towards its center becomes the main factor. Due to the Coriolis force action a spin-up of the converging fluid takes place in the direction of the general system rotation. In the turbulent convection regime in a non-rotating fluid there are fluctuations in both the velocity and temperature fields (see section 5.2). The value of the velocity of the fluid particle motion from the edges towards the center of the heated plate is a function of both the distance from the edge and the height over the edge. Due to any kind of asymmetry the fluid particle acquires a rotation component along the vertical symmetry axis and the rotation will depend strongly on the height over the heated plate. Therefore, above the surface at the values of the Rossby number $Ro \geq O(1)$ two conditions appear: an intensive vertical flux of heat and mass and an inhomogeneous differential rotation. These conditions are general for the origin of intense vortices of tornado or smaller scale.

Experiments on modelling intense atmospheric vortices were carried out by different methods. They were experiments in an annulus with a heated bottom and side walls with sloping plates forcing the warmed fluid to rotate near the walls in a certain direction and then this rotation was transferred to the whole volume of the fluid and in the system an inhomogeneous rotation of the whole rising fluid started. The other types of experiments use a jet injected into an already rotating fluid or a spined-up fluid introduced into the convection chamber. A detailed description of these experiments, having a centuries long history, can one find in the book by Bengtsson & Lighthill (1982).

The main results of these experimentations is the formation of intense solitary vortices. In most experiments a singular vortex was formed and its structure and stability in dependence on the external parameters was studied: the rate of spin-up and intensity of the rising flow. In other experiments, the rising flow was from a sufficiently large surface and several solitary vortices could arise and interact. A convective source is one of the methods of producing the rising flow, but not the only one. Also the general rotation of the system is not an obligatory condition. There were even more experiments in the system at rest in which an initial spin-up is created at the side walls. The general system rotation gives contradictory effects for modeling intense solitary vortices. On the one hand, the Coriolis force produces inhomogeneous rotation and intensifies the vortex production, but on the other hand, the general rotation weakens the horizontal converging motion from the vessel edges and produces several intense solitary vortices of the type observed in the turbulent regime of convection in the plane layer as discussed in Chapter 2.

At an inhomogeneous spin-down of the fluid such vortices were produced over the local heat sources (Boubnov & Heijst,1994); the main characteristics of such vortices are general enough. Some of the stages of such vortices are shown in Fig.79 where two main types of motion are seen: a spiraling vortex (Fig.79a) and a thorus-like motion (Fig.79b).

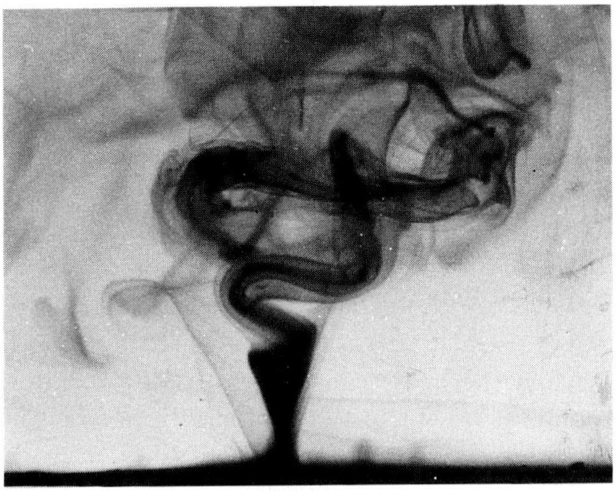

(a)

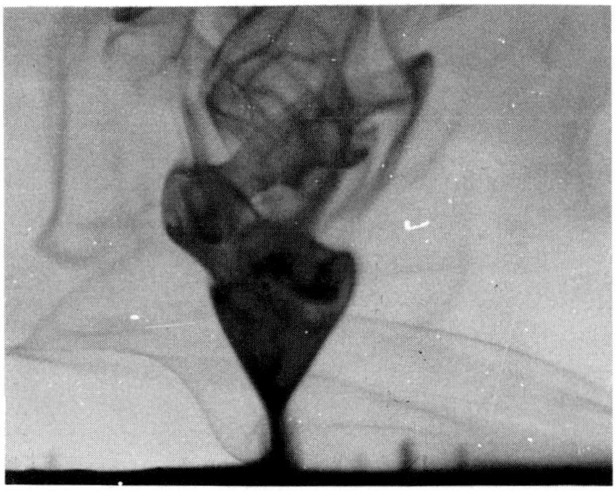

(b)

Figure 79: Examples of two stages of an intensive vortex: (a) - spiral, (b) - toroidal. (After Boubnov & Heijst, 1994).

The spiraling vortex is the basic type of a vortex characteristic for practically all experiments on modelling the solitary vortices for the free convection case (Fig.79) as well as for forced flows (injection of the fluid from the sides, jets into the rotating fluid or spinning-up jets into the fluid at rest, etc.). This spiraling vortex transports intensively the heat from the bottom. In dependence on how the transported heat is compensated, various types of spiraling vortices are possible: from a laminar one which is similar to vortices from the vortex grid in the rotating convective plane layer described in section 2.2, through more intensive conical spiraling vortices (see Fig.79a), to several spiraling vortices. The main problem for studies is the stability of such a vortex.

The free convection is a dynamically isolated system (Boubnov & Heijst, 1994) and the spiraling vortex in it can transfer the heat more intensively than it is supplied to the lower surface. As a result, such a vortex becomes unstable and instead a torus-like vortex flow a flow arises the main feature of which is the flow converging inside the torus (Fig.79b) when the heat transfer from the bottom decreases substantially. Then the bottom temperature starts to rise and after a time the spiraling vortex can form again and the process becomes periodical in time.

There is much in common for the various process under study. In the turbulent convection from a plane horizontal surface the periodical process in the layer was caused by a periodic breaking up and subsequent restoration of the thermal boundary layer (see section 2.6); at the convection from locally heated bodies the process was developing on the background of an intensive converging flow from the edges to the center and the characteristic frequency was revealing itself even more clearly (see section 5.1). Similar processes lead to periodicity in the regimes in the case of a solitary vortex. The essence of these processes is the periodic accumulation or decrease of heat in the thermal boundary layer: (it is with no regular motions as in the case of the Howard) when a cold film is formed at the free surface, in the presence of radial velocities for the case of a limited source or in the presence of both radial and azimuthal velocities in the case of intense solitary vortices. Similar processes are characteristic not only for convective processes. For instance, instabilities and breaking-up of spiral vortices is possible at the mechanical production of rotating jets and regular or coherent structures can form at distortions or breaking-up of boundaries in various hydrodynamic and physical processes.

Chapter 6
CENTRIFUGAL EFFECTS

In rotating systems centrifugal forces are always present. These forces are described by two additional terms in the momentum equation (1.1): by an addition to the pressure field $P = P' - \frac{1}{2}\left[\overline{\Omega} \times \overline{r}\right]^2$ and by the introduction into the external forces balance of the centrifugal force $\Omega^2\overline{r}$ where $\overline{r}$ is the distance from the rotation axis. The action of the force on fluid particles at different temperatures is equivalent to the gravity force action though their directions differ. The resulting buoyancy force will be determined as the vector sum of these two forces and their ratio is called the rotational Froude number: $Fr_\Omega = \frac{\Omega^2 L}{g}$. In the preceding chapters we neglected this force which is equivalent to $Fr_\Omega = 0$. For the plane layer this means that there is a basic state: the fluid at rest with the temperature gradient determined by the heat conduction. If there is a component of the external body force perpendicular to the temperature gradient then the fluid solid body rotation is impossible and the basic state will correspond to an axisymmetric circulation with the temperature field depending on the vertical and horizontal coordinates.

In this chapter we shall consider cases with very large Froude numbers, $Fr_\Omega >> 1$, and also some intermediate values, i.e. $Fr_\Omega \sim 1$. Difficulties of the consideration of the centrifugal effect in comparison with the Coriolis force are in its dependence on the distance from the rotation axis, i.e. the Froude number within the fluid is proportional to the distance $r = |\overline{r}|$. The asymptotic $Fr_\Omega >> 1$ assumes that fluid motions are considered at a distance from the axis. The simplest is a narrow cylindrical layer of a fast rotating fluid with which we shall start the consideration of the effects in section 6.1. Then the influence of the sloping end walls in such cylindrical layers is considered which substantially transforms the convective flow patterns (section 6.2). The joint consideration of these effects allows one to study motions in rotating spheres and spherical shells in Chapter 7 at the condition that the centrifugal force replaces the gravity force which models motions in a sphere with centrally symmetric gravitation. And finally, in section 6.3 we shall analyze convective motions in the plane horizontal layer with a finite lateral scale to see how the centrifugal force affects the motions described in Chapter 2.

6.1 Narrow cylindrical layer of a constant depth

Now we consider after Clever & Busse(1979) the stability of the fluid between two cylinders with the same central rotation axis. The outer cylinder has the temperature T_2 and the inner one T_1. The mean radius of the cylindrical shell is r_0 and the distance between the cylinders D^* is small compared to r_0, i.e. $D^* << r_0$. Such an assumption allows one to neglect the curvature of the walls and to introduce a Cartesian coordinate system. The depth of the layer, h, is assumed to be constant and the boundaries rigid. The condition of the constant depth is important because a varying depth leads to a deviation from the Taylor-Proudman theorem. The main attention shall we pay to a study of the friction in the Ekman boundary layers.

Using as the scales of length, time and temperature the values of h, Ω^{-1} and $\frac{(T_2-T_1)\,\Omega\, h^3}{D^*\, k}$, one can write down the Boussinesq equation system as

$$\frac{\partial \overline{V}}{\partial t} + \left(\overline{V}\nabla\right)\overline{V} = E\,\nabla^2\overline{V} - \nabla P - R\,E^{\frac{1}{2}}\left(\overline{i} - \overline{k}\,\frac{g}{\Omega^2 r_0}\right)\theta - 2\left[\overline{k}\times\overline{V}\right] \tag{6.1}$$

$$\nabla\overline{V} = 0 \tag{6.2}$$

$$\frac{\partial\theta}{\partial t} + \overline{V}\,\nabla\theta = \frac{k\,\nabla^2\theta}{\Omega\, h^2} \tag{6.3}$$

where the coordinate system x, y, z is connected to the cylinder axis and to the unit vectors $\overline{i}$, $\overline{j}$, $\overline{k}$; the vector $\overline{i}$ represents the direction of the centrifugal force. The non-dimensional numbers are: $E = 2\,Ek = \frac{\nu}{\Omega\, h^2}$, the doubled Ekman number, and

$$R = \frac{\alpha\,(T_2 - T_1)\,\Omega\, r_0\, h^2}{k\,E^{\frac{1}{2}}\,D^*} \tag{6.4}$$

which determines the balance between the buoyancy and viscous stress in the Ekman layer and is equivalent to the Rayleigh number but with the centrifugal acceleration $\Omega^2\, r_0$ instead of g and with $E^{\frac{1}{2}}\Omega\, h^{-2}$ instead of the kinematic viscosity. A fast rotation means that

$$Fr_\Omega = \frac{\Omega^2\, r_0}{g} >> E^{\frac{1}{2}} \quad \text{and} \quad \frac{\Omega^2\, h}{k} >> E^{\frac{1}{2}}$$

The equations (6.1)-(6.3) have an axisymmetric stationary solution identical to the "thermal wind" relationship:

$$\overline{V} = \overline{U} = \overline{j}\,(z + c')\,R\,E^{\frac{1}{2}}\,\frac{g}{\Omega^2\, r_0}\,\frac{k}{2\Omega\, h^2} \tag{6.5}$$

$$\theta = \Theta = x\,\frac{k}{\Omega\, h^2} + \frac{T_2 - T_1}{2\,(T_2 + T_1)} \tag{6.6}$$

and the boundary condition of the thermo-insulation of the upper and lower boundaries are

$$\theta - \frac{T_2-T_1}{2(T_2+T_1)} = \pm\frac{D^* k}{2\Omega\, h^3} \quad \text{at} \quad x = \pm\frac{D^*}{2h}$$
$$\frac{\partial\theta}{\partial z} = 0 \quad \text{at} \quad z = \pm\frac{1}{2} \tag{6.7}$$

The solution corresponding to the "thermal wind" is fulfilled within the fluid but the boundary conditions for zero velocity at the boundaries is not fulfilled. At large rotation rates, the boundary layers near the surfaces are thin and in the limit $R\,E\,D^*\,g\,\left(h\,\Omega^2\,r_0\right)^{-1} << 1$ one can suppose that the solution is fulfilled everywhere within the fluid. The integration constant c' in Eq.(6.5) is zero for the rigid upper surface and $c' = \frac{1}{2}$ for the free one. To study the stability of this basic state small perturbations are considered for the velocity $\overline{V} = \overline{U} + \overline{V'}$ and temperature $\theta = \Theta + \theta'$ and the equations for the perturbations $\overline{V'}$ and θ' are analyzed; they are (the sign $'$ is dropped further on):

$$\frac{\partial\overline{V}}{\partial t} + \left(\overline{U}\nabla\right)\overline{V} = E\,\nabla^2\overline{V} - \nabla P - R\,E^{\frac{1}{2}}\left(\overline{i} - \overline{k}\,Fr_{\Omega}^{-1}\right)\theta - 2\left[\overline{k}\times\overline{V}\right]$$
$$\frac{\partial\theta}{\partial t} + \overline{U}\,\nabla\theta = \frac{k}{\Omega\,h^2}\left(\nabla^2\theta - \overline{i}\,\overline{V}\right) \tag{6.8}$$

Considering the fast rotation, one can use as a small parameter the value of $E^{\frac{1}{2}}$ and can expand $\overline{V}$ and θ on it:

$$\overline{V} = \overline{V}_0 + E^{\frac{1}{2}}\,\overline{V}_1 + \dots\,, \qquad \theta = \theta_0 + E^{\frac{1}{2}}\,\theta_1 + \dots \tag{6.9}$$

As small terms we include into the expansions additions in the viscous boundary layers whose thickness at the boundaries parallel to the rotation axis is of order $E^{\frac{1}{3}}$, the so-called Stewartson layers (see Table 1 in Chapter 2). Then assuming that the energy dissipation within the main fluid volume is small comparing to dissipation in the Ekman layers, the boundary conditions at the side walls can be presented as

$$\left(\overline{n}\,\overline{V}\right) = 0 \tag{6.10}$$

where $\overline{n}$ is the unit vector perpendicular to the boundary surface. We assume that the velocity time derivative is small compared to unity and can be expanded as $\exp\left(E^{\frac{1}{2}}\,\sigma\,t\right)$. Substituting expansions (6.9) into the first Eq.(6.8), we obtain the basic balance as

$$2\left[\overline{k}\times\overline{V}_0\right] + \nabla P_0 = 0, \qquad \nabla\overline{V}_o = 0 \tag{6.11}$$

The general solution of these equations satisfying the condition $\overline{k}\cdot\overline{V}_0 = 0$ at $z = \pm\frac{1}{2}$ is

$$\overline{V}_0 = \frac{1}{2}\left[\overline{k}\times\nabla P_2\right] \tag{6.12}$$

where P_2 is an arbitrary function of x and y which does not depend on z. The boundary conditions (6.10) require the value of P_2 at $x = \pm\frac{D^*}{2h}$.

Considering the corresponding equations for $\overline{V}_1$ for the case $\sigma = 0$, one can show that the equation for the function P_2 can be written as

$$\Delta^2 P_2 = -\frac{R}{2}\frac{\partial^2 P_2}{\partial y^2} \tag{6.13}$$

where, as usual, $\Delta = \left(\frac{\partial^2}{\partial x^2}\right) + \left(\frac{\partial^2}{\partial y^2}\right)$. The boundary conditions $P_2 = \frac{\partial^2 P_2}{\partial x^2} = 0$ at $x = \pm\frac{D^*}{2h}$ are satisfied for the following solution of Eq. (6.13):

$$P_2 = \exp(iay)\cos\left(\frac{\pi x h}{D^*}\right) \tag{6.14}$$

with the eigen value

$$R = \frac{2\left[\left(\frac{\pi h}{D^*}\right)^2 + a^2\right]^2}{a^2} \tag{6.15}$$

The critical value of the Rayleigh number R_{cr} is determined as the minimum of the function $R(a)$ and is equal to

$$R_{cr} = 8\left(\frac{\pi h}{D^*}\right)^2, \qquad a_{cr} = \frac{\pi h}{D^*} \tag{6.16}$$

When $R = R_{cr}$, the symmetric state (6.5) becomes unstable. Note that this conclusion implies that $\frac{h}{D^*}$ is of order unity or that the inequalities

$$E^{\frac{1}{2}} << \frac{h}{D^*} << E^{-\frac{1}{2}} \tag{6.17}$$

are valid. The left part of it is related to the applicability of the boundary layer method. For very thick layers the friction can be neglected in the Ekman layers but then the right part of the inequality will not be satisfied. In this case the problem becomes mathematically identical to the Benard convection in a boundary layer. Here the gravitation is replaced by the centrifugal force and the Coriolis force does not influence the convection origin. The usual Rayleigh number here is $Ra = Ra_{cr} = 1708$ and taking into account the relationship between Ra and R as determined by Eq.(6.4), we have

$$Ra = \frac{R}{E^{\frac{1}{2}}}\left(\frac{D^*}{h}\right)^4 = 1708 \tag{6.18}$$

Results of experimental tests of the convection onset at the centrifugal force action have been presented by Busse & Carrigan (1974). The rotation rates there were up to 400 revolutions per minute which allowed to reach centrifugal accelerations up to 7g. After the convection onset, rolls appeared with axes parallel to the rotation axis which did not move relatively to the cylinder wall.

The first experiments were carried out for large values of $\frac{h}{D*} \approx 48$. For the value of E^{-1} in the range from $4 \cdot 10^4$ to $3 \cdot 10^5$, the theoretical result represented by Eq.(6.18) is fulfilled well in the experiments which confirms the assumption that the main dissipation takes place within the layer and the critical Rayleigh number does not depend on the rotation rate of the system (if $g \leftrightarrow \Omega \cdot r_0^2$). Some deviations are observed for $E^{-1} \leq 7 \cdot 10^4$ which may be explained by baroclinic effects.

Investigations of the convection onset at the main dissipation taking place in the Ekman boundary layers were carried out at $\frac{h}{D*} = 0.97$ and 0.46. The experimental value of R_{cr} was almost twice of that given by Eq.(6.16). This discrepancy was decreasing with the value of $\frac{h}{D*}$; however, a further decrease of this value below 0.46 was not allowed by the experimental equipment.

6.2 Layer with sloping boundaries

At the derivation of the conditions for the convection onset in the preceding section 6.1, it was assumed that $\sigma = 0$, i.e. the stationary solution and the main balance were geostrophic when the Coriolis force was equilibrated by the pressure gradient. The Taylor-Proudman theorem allows deviations of higher order from two-dimensional perturbations; therefore, at the sloping end layer boundaries the horizontal scale of the critical perturbations should decrease in order to minimize the influence of the depth changes. For the sloping end boundaries, the geostrophic mode will not be a stationary mode any more and the imaginary part of σ does not vanish in the critical perturbations. This means that the rolls move with a constant angular velocity relative to the cylinder walls. Such a mode of motion is sometimes called the "thermal Rossby waves".

Following Busse & Or (1986), now we consider the rotation of a cylindrical layer with the upper and lower parts of the layer inclined at some small angle to the horizon. This means that the unit normal vector at the top and the bottom of the layer, $\overline{n}_t$ and $\overline{n}_b$, deviate at a small angle from the direction of the rotation axis. The value $\eta = \frac{1}{2}\overline{i}\,(\overline{n}_t + \overline{n}_b)$ is introduced and it is proportional to the layer depth change. The dissipation is taken into account only in the Ekman layers. The solution for the zero mode is, as before, described by Eq.(6.11). To the boundary condition for the velocity component normal to the walls $(\overline{n} \cdot \overline{V}_1)$ the term - $\frac{1}{2}\left[\overline{n} \times \overline{k}\right] \nabla P_2$ is added and instead of Eq. (6.13) we shall have the following equation:

$$\left(\triangle - i\lambda\, Pr\, E^{-\frac{1}{2}}\right)\left[\triangle\,(2 + i\lambda) - 4\eta\, E^{-\frac{1}{2}}\left(\frac{\partial}{\partial y}\right)\right] P_2 = -R\,\frac{\partial^2 P_2}{\partial y^2} \tag{6.19}$$

where $\sigma = i\lambda$. The solution is sought in the form (6.14) and then the dispersion relation for λ is obtained:

$$\lambda\left[2Pr\, E^{-\frac{1}{2}} + \left(\frac{\pi h}{D*}\right)^2 + a^2\right] = -4\eta\, E^{-\frac{1}{2}} \cdot a \tag{6.20}$$

The real part of Eq.(6.19) taking account of (6.20), gives an expression for the eigenvalues of R:

$$\frac{R}{2} = \left[\left(\frac{\pi h}{D^*}\right)^2 + a^2\right] + \left(\frac{4\pi\, Pr}{E}\right)^2 \left[2Pr\, E^{-\frac{1}{2}} + \left(\frac{\pi h}{D^*}\right)^2 + a^2\right]^{-2} \tag{6.21}$$

Determination of the minimum of this function $R = R(a)$ gives the critical values of $R = R_{cr}$ and $a = a_{cr}$ which at the condition

$$\eta << E^{\frac{1}{4}}\, Pr^{\frac{1}{2}} \tag{6.22}$$

are of the same order as for the case of the constant depth layer and the critical values of a are of order unity.

A more interesting case is an increase of the slope η and a consideration of the dissipation in the total layer volume. The dissipation in the Ekman layers is negligible if $a >> E^{-\frac{1}{2}}$, or, taking into account the expression for a in Eq. (2.27), we shall have

$$1 >> \eta\, Pr >> E^{\frac{1}{4}} \tag{6.23}$$

Then in Eq.(6.9), instead of $E^{\frac{1}{2}}$ as a small parameter, we shall use the value of η. Then for the equation for the eigenvalues, assuming the dissipation only in the internal regions of the layer, we obtain instead of Eq. (6.13)

$$\left(\Delta - \frac{i\lambda'\eta\, Pr}{E}\right)\left[\frac{E\,(\Delta - i\lambda')}{\eta}\Delta + 4\left(\frac{\partial}{\partial y}\right)\right] P_0 = R^* \left(\frac{\partial^2 P_0}{\partial y^2}\right) \tag{6.24}$$

where the perturbations are supposed to be of the form $\exp\,(i\lambda'\eta t + iay)$ and $R^* = R\,E^{\frac{1}{2}}\,\eta^{-1}$. Assuming that all changes along x are small in comparison with changes along y from the imaginary part of (6.24), we obtain the dispersion relation:

$$(1 + Pr)\, a\lambda' + 4 = 0 \tag{6.25}$$

and the real part of (6.24) gives a corresponding expression for R^*:

$$R^* = \frac{a^4\, E}{\eta} + \frac{16 Pr^{\,2}\, \eta}{E\, a^2\, (1 + Pr)^2} \tag{6.26}$$

If one drops the factor $(1 + Pr)$ from (6.25), it would become the dispersion relationship for the usual Rossby waves. The critical values for the critical values of the Rayleigh-type number and of the wave number are:

$$a_{cr} = \left[\frac{2\sqrt{2}\,\eta\, Pr}{E\,(1 + Pr)}\right]^{\frac{1}{3}} \tag{6.27}$$

$$R^*_{cr} = 3\left(\frac{\eta}{4E}\right)^{\frac{1}{3}}\left(\frac{4Pr}{1+Pr}\right)^{\frac{4}{3}} \tag{6.28}$$

One of the features of the expression of (6.28) for the critical value of R^* is its dependence on the layer thickness D^*.

Comparison of this theoretical results with the laboratory experiments studying transition from the lower axisymmetric regime to the wave regime (see Fowlis & Hide, 1965) in a rotating annulus with the free surface (which is always slightly inclined to the horizon) showed some overestimate of the critical parameter values for the instability of the symmetric circulation.

Experiments on the model test carried out by Busse & Carrigan (1974) confirmed its main conclusions. A good correspondence of the results was observed for slopes $\eta \leq 45°$, for the dependence of the critical value of R^* on E^{-1} changing from $6 \cdot 10^3$ to $2 \cdot 10^4$. This is astonishing because the theory was developed for small values of η and 45° corresponds in radians to 0.8. Also fairly good was the confirmation of the independence of the critical value of R^* on the gap thickness D^* while it was changing in the range of 60 per cent.

An expansion of these experiments to $\eta = 47°$ and using two liquids with widely different properties (water, $Pr \approx 7$ and mercury, $Pr = 0.025$) was devoted mainly to the determination of the r.m.s. temperature fluctuations on the value of ordinary Rayleigh number in which the acceleration g was replaced by $\Omega^2 r$ as in Eq.(6.18). The main result of this work is that the temperature fluctuation frequency increases with the Rayleigh number and that there is a maximum of the r.m.s. fluctuation value at the value of the Rayleigh number which depends only weakly on the values of the parameter E and the Prandtl number. Also studied was the dependence of the prevailing perturbation wave length on the Rayleigh number which showed that this length changes weakly with E.

The sloping boundaries of the cylindrical layer considered above were plane. If the layer is confined by surfaces which are convex or concave then a perturbation analysis can be carried out (see Busse,1982) in which the value of curvature is used as the small parameter. The main conclusion of this analysis is that the velocity of the thermal Rossby waves is a function of the distance from the central axis. For the convex surfaces a zonal shear flow arises when the motion near the internal cylinder is directed against the general rotation of the system near the outer cylinder and the flow overtakes the general rotation. In case of concave bounding surfaces, the shear flow changes to the opposite.

Or & Busse (1987) and Schnaubelt & Busse (1992) used the two-dimensional model for convection with the sloping boundaries. In the frame of this model a large variety of different states of motion such as vacillation and spatially modulated flow was found. Here the term vacillation means variation of spatially averaged properties in time as well as spatially fluctuating fields. The stability diagram was plotted for different regimes depending on the values of the Rayleigh and Prandtl numbers and the Coriolis parameter.

6.3 Shallow cylindrical layers

Experimentally, the onset and patterns of convective flows in a infinite plane horizontal layer were usually studied in a shallow cylinder vessel with a diameter, $2L$, substantially larger than the layer depth h, i.e. $\frac{L}{h} >> 1$. Therefore, all the experimental results described in Chapter 2 are influenced by the centrifugal forces. In a general case of comparison of experimental and theoretical results, it is very difficult to separate the role of the centrifugal effects from the non-linear ones (finite amplitude convection) or from the influence of the side walls, and all these effects are often interrelated. Therefore in this paragraph we shall consider only some results of asymptotic methods for the cases when the rotational Froude number is not small and centrifugal forces either prevail or are comparable to other forces.

The history of the problem development beginning from the studies of centrifugal convection in rotating turbines can be found in a paper of Homsy & Hadson (1969). Of the more important results we note the existence of radial flow in the viscous boundary layers of Ekman thickness of order $\left(\frac{\nu}{\Omega}\right)^{\frac{1}{2}}$ directed to the center of a solitary disk in the rotating fluid when the temperature of the disk is increasing: the so-called thermal spin-up (see Daniels,1980). When the temperature of the disk is decreasing, a stationary motion is possible whose perturbations at small temperature changes show the presence of a viscous sublayer with a thickness less than for thermal boundary layers. For large Prandtl numbers, computations have been carried out for the case when thicknesses of the viscous and thermal boundary layers are of the same order. The case of two infinite coaxial discs was considered. The flow to the center of the disk was observed in the upper Ekman layer and the flow from the center in the lower layer. It was shown that there is a critical parameter $\Lambda = Pr\, Ro_\alpha\, Ta^{\frac{1}{4}}$, characterizing the convection efficiency relative to the heat conduction and determining the solution of the problem, where $Ro_\alpha = \frac{1}{4}\alpha T$ is the thermal Rossby number. An asymptotic solution has been found for the case $\Lambda >> 1$ and $Ta >> 1$ in which near the disks the thermal boundary layers form with thickness of the order $\left(Ta^{\frac{1}{4}}\, Pr\, Ro_\alpha\right)^{-1} = (\Lambda)^{-1}$. A large heat transfer existed at the lower disk while at the upper one it was equal to zero.

Homsy & Hadson (1969), for a large range of this parameters, have carried out an analysis of the centrifugal force influence on convection at fast rotation by simultaneous accounting for both centrifugal and Coriolis forces. The upper surface is homogeneously heated at thermo-insulated (or perfectly conducting) side walls of a cylindrical vessel with radius L. Without centrifugal forces such a system is stable in the Boussinesq approximation. In papers of Homsy & Hadson (1971a,b), the linear stability analysis was expanded to the case of the heating from below. The stability of convective motions was considered in the ranges: $Ro_\alpha << 1$, $\Lambda L \le O(1)$ and $\Lambda L << Fr << E^{\frac{1}{3}}$ for axisymmetric perturbations and $\Lambda L << Fr << E^{-\frac{1}{3}} \frac{L}{m}$ for non-axisymmetric perturbations with the azimuthal wave number m. Here $E = 2\, Ek$ as in section 6.1.

For the case of fast rotation, i.e. $E << 1$, one may use the boundary layer approximation, i.e. the circulation motion in the Ekman layers and the gravity instability are mainly caused by a forces disbalance at some distance from the horizontal boundaries. The influence of the centrifugal forces in the linear analysis was taken into account by changing the temperature profile while in the equations for the perturbations the forces were absent.

Torrest & Hadson (1974) solved the linear problem by the Galerkin method and in the range of the parameters considered the equations do not have terms describing deviations from the fluid solid rotation and centrifugal forces. The influence of the latter is not found at $\Lambda \leq 1$ and the solution coincides with the one for the infinite layer without these forces. At increasing the value of Λ ($L > 1$ and $1 \leq \Lambda \leq 10$ were considered), the centrifugal forces stabilizes the convection and the stabilizing effect monotonically grows with Λ and L.

Daniels (1980) studied the influence of centrifugal forces on the axisymmetric motions in a plane layer for the case of stationary convection when the flow due to the vertical gradient is comparable with the flow due to forces. The stratification in all directions was taken into account and therefore the basic system of equations differed significantly from the usual Boussinesq approximation. The horizontal walls had perfect heat conduction while the vertical ones were perfectly thermo-insulated. The solution was considered near the critical values of the Rayleigh number corresponding to an infinitely long fluid layer and the expansion was carried out on the small parameter $\delta = \frac{h}{L} << 1$. It was shown that for both free or rigid boundaries the centrifugal acceleration could give a double effect: a possibility of a smooth transition to the finite amplitude convection represented by non-zero boundary conditions for $\overline{V}$, i.e. $\overline{V} \neq 0$, and determined by the amplitude function A_f which may be compared to a treshold transition and the existence of critical values of the Rayleigh number in the case of no centrifugal forces. The second effect is the presence of non-linear interactions of the vertical instability with the centrifugal forces described by the appearance of a new term $C_A\, A_f\, r^2\, L^{-2}$ in the amplitude equation, where r is the distance from the center and C_A is a constant, positive or negative in dependence of the values of the determining parameters. For large values of the Taylor number in dependence of the sign of the value C_A, the effective Rayleigh number corresponding to the onset of the finite amplitude convection can be both larger or smaller than the critical Rayleigh number for the limited layer. For the Prandtl number $Pr \geq 1$ at large values of the Taylor number, the value of C_A is positive and there is no subcritical bifurcation. Convection cells appear for substantially smaller values of Ra_{cr} than for the infinite layer and their amplitude grows with the distance from the center. A certain critical radius exists bounding the circle inside of which there is no motion. With the Rayleigh number, increasing the critical radius tends to zero and for $Ra \to \infty$ the intensity of the cells is homogeneous.

For $C_A < 0$ and the Prandtl number of order unity, the convection motion, vice versa, is developed inside the region bounded by a critical radius though the motion at the outer vertical wall caused by the centrifugal force is also present. These results are a possible explanation of the appearance of the first ring roll near the side wall in the experiments by Koschmider (1966,1968) at values of the Rayleigh number much less than its critical value for the infinite layer.

Chapter 7
CONVECTION IN SPHERES AND SPHERICAL SHELLS

Many problems of motions in stars, planetary atmospheres and cores require description of convection in spherical shells and within spheres which may be considered as a partial case of a sphere with zero internal radius. A spherical geometry with a central gravity field is pertinent to natural objects. In the laboratory, experiments on the Earth's surface are in a homogeneous gravity field. For the spheres rotating in laboratory conditions, we have a superposition of two types of symmetry: one is related to the spherical geometry and the other (cylindrical) is related to the rotation. This superposition complicates the theoretical consideration.

Beside theoretical complications, there are serious difficulties in conducting proper laboratory experiments. Checking results for the self-gravitating spheres requires force fields with spherical symmetry and avoiding the Earth gravity field. This is much more complicated in comparison with a plane horizontal layer or coaxial cylinders and one of the experiments was conducted on the Space Shuttle to avoid the Earth gravity and its central forces were of electric origin.

Nevertheless, considerable attention has been paid recently to theoretical analysis and laboratory studies of convective motions in spherical layers because for many geophysical and astrophysical problems considerations of motion are impossible without taking into account the effects of sphericity. In this chapter we consider some general aspects of the problem. In more details the problem from the motion stability point of view, was considered by Chandrasekhar (1961) and the evolution of the studies by Busse (1982) and Yavorskaya & Belyaev (1982).

7.1 Spherically symmetric gravitation and heating

For convective motions in spherical shells and within a sphere, the question on the type of symmetry is one of the determining, because at the transition to consideration of motions in separate parts of the spherical shell (near the equator, or near the poles) one can asymptotically move to a simpler case of convection: to the plane layer with parallel or slightly inclined boundaries with rotation parallel or perpendicular to the direction of the gravity force. These two cases can be schematically presented after Busse (1982), Busse & Cuong (1982) as two plane layers rotating with a constant angular velocity Ω around an axis parallel (Fig.80a) or perpendicular (Fig.80b) to the layer boundaries.

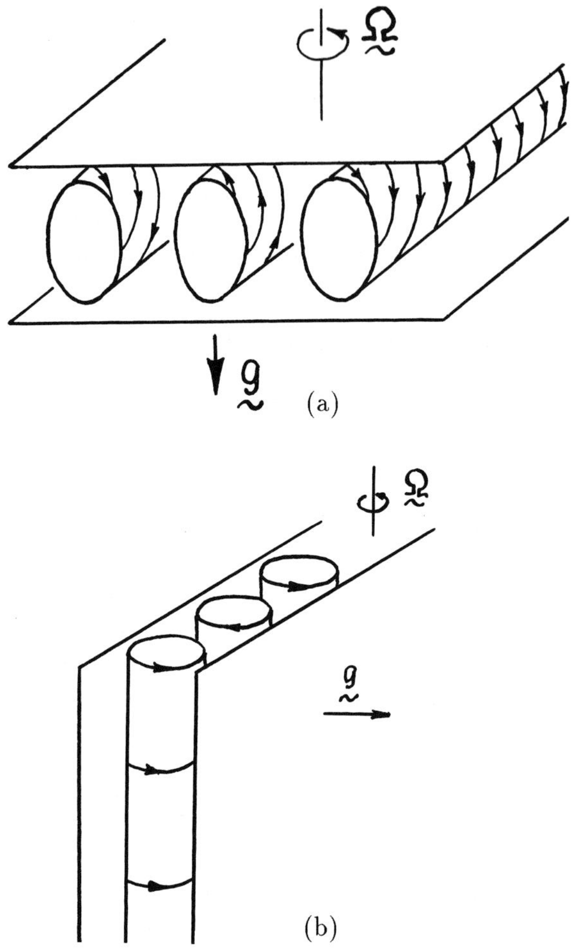

Figure 80: Two main cases for convection in a layer heated from below and rotating: (a) about a horizontal axis of rotation (the motion shown here satisfies the Proudman-Taylor theorem) (b): about a vertical rotation axis. (After Busse, 1986, courtesy of Cambridge University Press).

The first case corresponds to regions near the equator and was considered in detail in section 6.1, the second case corresponds to the polar regions. The origin and development of convection for the second case was considered in Chapter 2 and the influence of the centrifugal forces in section 6.3. The first case may also be considered as a limiting case of convection in a cylindrical vessel, i.e. in an annulus. The main difference of convective and baroclinic instability in the annulus is in the directions of the determining vectors: the gravity acceleration vector, $\overline{g}$, and the temperature gradient, ∇T. For the baroclinic instability case $\overline{g} \perp \nabla T$ and in the main fluid core

$(\overline{g}\,\nabla T) < 0$. If this scalar product is positive, the case differs slightly from the case $(\overline{g}\,\nabla T) = 0$, or the purely convective case.

A transition from the case of the convective instability in annuli to an analysis of the motions on a sphere can be carried out taking into account the influence of sloping boundaries on the onset and structure of the convective motions (section 6.2). Using the results of sections 6.2 and 6.3, we will consider now the conditions for convection onset in the problem with a spherically symmetrical heat source and body force, the case characteristic for many geophysical and astrophysical problems.

Following Busse (1970), we shall now consider motions in a sphere of radius r_0 assuming that the heat and density sources distribution is such that an unperturbed solution is possible with the temperature and pressure gradients given by

$$\nabla T = -\beta\,\overline{r} \tag{7.1}$$

$$\nabla P = -\rho_0\, g\,\overline{r} \tag{7.2}$$

where $\overline{r}$ is the vector from the center of the sphere and β, ρ_0, and g are constants. Using r_0 and $\beta r_0^3 \Omega k^{-1}$ as the scales of length and temperature, we can obtain equations for the velocity vector $\overline{V}$ and θ, the temperature deviation from the equilibrium distribution (7.1):

$$\frac{\partial \overline{V}}{\partial t} + 2\left[\overline{k} \times \overline{V}\right] = E\,\nabla^2 \overline{V} - \nabla P + Ra_s\,\overline{r}\,\theta \tag{7.3}$$

$$\nabla \overline{V} = 0 \tag{7.4}$$

$$\frac{\partial \theta}{\partial t} = \frac{k}{\Omega\, r^2}\left[\nabla^2 \theta + \left(\overline{r}\,\overline{V}\right)\right] \tag{7.5}$$

where the Rayleigh number for the sphere Ra_s is determined as

$$Ra_s = \frac{\alpha\,\beta\, g\, r_0^2}{k\,\Omega} \tag{7.6}$$

and related to the number R^* of section 6.2 as

$$Ra_s = R^* \left(\frac{\Omega^2\, r_0}{2g}\right) \tag{7.7}$$

Changes of the fluid depth in the direction of the rotation axis, the body force and the main temperature gradient depend on the distance from the axis. The approximation of the gap requires that instability scales should be small compared to unity as it was in section 6.2. According to Eq.(6.28) and (7.1), the critical value of Ra_s depends on depth changes in the power exponent 4/3. This change is slower with distance than the change of the buoyancy force at small distances from the axis and faster at large distances; therefore, one can expect that the smallest value of the

Rayleigh number will be at some finite distance from the axis. Denoting by $\sin\varsigma$ the distance from axis, the Eq.(6.24) taking account of (6.25), can be written as

$$\left(\Delta_2 - i\lambda_s \frac{Pr}{E}\right)\left[(E\,\Delta_2 - i\lambda_s)\,\Delta_s + 4\left(\frac{\sin\varsigma}{2\cos\varsigma}\right)\left(\frac{\partial}{\partial y}\right)\right] P_2 = Ra_s \sin^2\varsigma\left(\frac{\partial^2 P_2}{\partial y^2}\right) \tag{7.8}$$

Here, the fact was taken into account that the depth of the coaxial cylinders in the sphere is equal to $2\cos\varsigma$ instead of 1 and that the area of the surface is greater than its projection to the plane Oxy by a factor $\cos^{-1}\varsigma$. The expansion for P_2 is supposed to be of the form

$$P_2 = f(x)\exp\left[i\,(ay + \lambda^* t)\right] \tag{7.9}$$

Neglecting in (7.8) the dependence on the x coordinate for the sphere, one can rewrite the expressions for the critical values of the parameters Eqs.(6.25), (6.27) and (6.28) as

$$a_0 = \left[\sqrt{2}\,\frac{\sin\varsigma}{\cos^2\varsigma}\,\frac{Pr}{E\,(1+Pr)}\right]^{\frac{1}{3}} \tag{7.10}$$

$$Ra_{s0} = \frac{3}{E^{\frac{1}{3}}\sin^2\varsigma}\left[\sqrt{2}\,\frac{\sin\varsigma}{\cos^2\varsigma}\,\frac{Pr}{(1+Pr)}\right]^{\frac{4}{3}} \tag{7.11}$$

$$\lambda^* = \left[4\sqrt{2}\cdot\frac{\sin^2\varsigma}{\cos^4\varsigma}\,\frac{E}{Pr\cdot(1+Pr)^2}\right]^{\frac{1}{3}} \tag{7.12}$$

Index 0 instead of cr is used here because the critical values are determined for the minimum of the Rayleigh number from (6.29) as a function of ς. This minimum corresponds to the critical angle $\varsigma = 26.6°$. Therefore, the instability first arises in the cylindrical layer crossing the sphere at the latitude 63.4°. Fig. 81 presents a pattern of motions at the convection onset. The critical values of the parameters described by Eqs.(7.10)-(7.12) then are:

$$a_{cr} = \left[\sqrt{\frac{5}{8}}\,\frac{Pr}{E\,(1+Pr)}\right]^{\frac{1}{3}} \tag{7.13}$$

$$Ra_{s\,cr} = \frac{3}{E^{\frac{1}{3}}}\left(\frac{5}{4}\right)^{\frac{5}{3}}\left[\frac{2Pr}{(1+Pr)}\right]^{\frac{4}{3}} \tag{7.14}$$

$$\lambda^*_{cr} = -\sqrt{2}\left[\frac{5\cdot E}{Pr\,(1+Pr)^2}\right]^{\frac{1}{3}} \tag{7.15}$$

The change of the height at the critical latitude is not too small as compared to unity but the difference from the exact solution is only of the order of few per cent. An account of the depth change leads to the conclusion that it is permissible to neglect the velocity component parallel to the rotation axis.

Soward (1978) obtained a solution of the non-linear equations for the convection onset in a sphere with internal heating. It was shown that the critical values of the Rayleigh number differ from the ones given by Eq.(7.14) by a value of order $O(Ta^{-\frac{1}{9}})$ or $O(E^{\frac{2}{9}})$. For his consideration he took three characteristic linear scales: height of the cylindrical layer $O(1)$, the scale in the radial direction $O(Ta^{-\frac{1}{9}})$, corresponding to the width of the layer and a still smaller scale in the azimuthal direction corresponding to the perturbation wave length. It was shown that an intensive convection is possible in a layer of width $O(Ta^{-\frac{1}{18}})$ where solutions exist, called by the author "thermal solitary waves" drifting in the radial direction with a variable velocity. As the solitary wave approaches the boundary of the layer which is $O(Ta^{-\frac{1}{18}})$ wide, the conditions supporting convection are weakening and the waves degenerate.

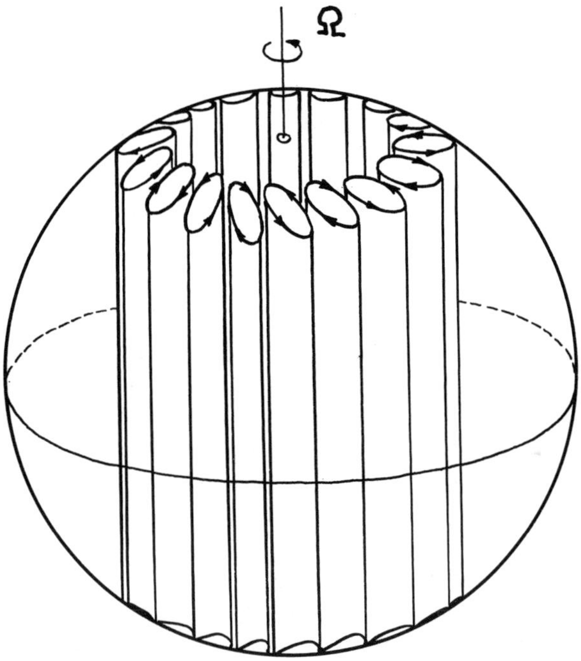

Figure 81: Sketch of motions at the onset of convection in a homogeneously rotating self gravitating sphere. (After Busse, 1970, courtesy of Cambridge University Press).

Studies of the amplitude convection in rotating spheres and spherical shells were performed mainly numerically. A review of the works in this direction can be found in the work by Yavorskaya & Belyaev (1982).

An experimental study carried out by Carrigan & Busse (1983) showed that convection in a wide spherical shell arises, as had been forecasted by the asymptotic theory in the form of a cylindrical layer consisting of convective columns. The experimental dependence of the critical Rayleigh number on the Taylor, or Ekman, number agrees qualitatively well with the theory. The quantitative difference lies in the value of the proportionality coefficient which is about twice its theoretical value for practically all values of the shell width. The centrifugal forces in these experiments were substantially larger than the gravity force and determined the direction of the buoyancy force action.

7.2 Thin spherical shells

The problem of the study of the convective motion in a rotating spherical shell is too complicated in the general case due to the impossibility of separation of the variables. Only the azimuthal dependence can be separated. Therefore, most of the studies have been performed for spherical layer and for various boundary conditions. The rotation as compared to the case of a sphere at rest leads to the circumstance that the value of the critical Rayleigh number corresponds to a certain value of the azimuthal wave number. The convection pattern is reminiscent of the one for the Rossby waves and consists of banana-shaped cells stretched from pole to pole with the amplitude maximum near the equator.

The main results of the linear theory are as follows:

(i) The most unstable modes are axially asymmetric though they are symmetric relative to the equator. The critical wave number of the most unstable mode grows with the rotation rate for $Ta > 10^5$ and its amplitude is maximal near the equator.

(ii) With the angular velocity increasing for $Ta > 10^3$, the modes are separated in two classes: equatorial modes with large wave numbers and polar ones.. With Ω increasing, the number of polar modes is decreasing and that of equatorial modes increasing.

(iii) All modes propagate azimuthally with the phase velocity ω. For $Ta \leq 10^3$ the modes propagate in the direction opposite to the rotation and for larger rates of rotation $Ta > 10^4$ the waves propagate in the direction of rotation. For the same value of the Taylor number, the modes with the small wave numbers have lesser velocities of propagation than modes with large wave numbers. The most unstable modes have a phase velocity close to the maximal one.

(iv) With the rotation rate increasing, the convective patterns change their shape. For small values of the Taylor number near the stability treshold the cells have a banana-like forms and for large values of the Taylor number, a transition is observed to cylindrical rolls with the vertical axis crossing the upper spherical boundary at some latitude.

The consideration of convection in section 7.1 for a sphere as an analog of the convection in a cylindrical layer with inclined walls is suited for studies of the convection in spherical shells for which the non-dimensional distance from the rotation axis is larger than the width of the gap $\eta^* = 1 - \frac{r_i}{r_o}$, where r_i is the radius of the internal sphere. For applications to the geophysical fluid dynamics problems this means a considerations of regions away from the poles.

In the case of polar regions $0 < s < \eta^*$, one should consider a plane infinite layer with inclined boundaries as in section 6.2. An analysis of perturbations carried out by Busse & Cuong (1977) shows that such boundaries cause a drift of convective rolls. This drift is in the meridional direction, from the rotation axis, and intensifies with increasing thickness of the layer if $Pr > \frac{1}{3}$. Critical values for the case of a small gap for $Pr > 0.677$ are

$$Ra_{g\,m} = 3\left[\frac{2\pi^2 E \cdot (1-s^2)}{(1-\eta^2)}\right]^{\frac{2}{3}} \tag{7.16}$$

$$a_m = \left[\frac{4\pi\,(1-s^2)^{\frac{1}{2}}}{E\,(1-\eta)}\right]^{\frac{1}{3}} \tag{7.17}$$

$$\omega_m = -\frac{2\,s}{(3Pr-1)\;a_m \cdot (1-s^2)^{\frac{1}{2}}} \tag{7.18}$$

where

$$Ra_g = \frac{\alpha\, g\, \beta\, \nu\, r_0}{k\Omega^2}$$

For the Prandtl numbers $Pr < 0.677$, the values of the critical parameters can be obtained from the Chandrasekhar (1961) asymptotic theory:

$$Ra_{g\,m} = 6\left[\frac{2\pi^2 Pr^{\,2}\, E\,((1-s^2)}{(1+Pr)^{\frac{1}{2}}\;(1-\eta)^2}\right]^{\frac{2}{3}} \tag{7.19}$$

$$a_m = \left[\frac{4\pi\,(1-s^2)^{\frac{1}{2}}}{E\,(1+Pr)\,(1-\eta)}\right]^{\frac{1}{3}} \tag{7.20}$$

$$\omega_m = -\frac{2\pi\left(1-\frac{3Pr^{\,2}}{2}\right)^{\frac{1}{2}}\,(1-s^2)^{\frac{1}{2}}}{(1+Pr)\;a_m\;(1-\eta)} \tag{7.21}$$

If one compares results for the convection onset before and after the critical value of the latitude $s = \eta$, then one finds that at the cylindrical surface there is a discontinuity and the critical Rayleigh number drops fourfold at this transition. The pattern of the convective mode changes at this value: at $s > \eta$ it resembles the convection in a plane layer rotating around the vertical axis while at $s < \eta$ it resembles the convection in a cylindrical layer in the form of vertical columns tangent to the equator of the internal sphere at $\eta > 0.5$ and $s = \eta$. The Rayleigh number reaches its absolute minimum always at $s > \eta$.

Experimental studies of the convection in a rotating spherical shell are rare. In a paper of Carrigan & Busse (1983) the experiments were carried out for various values of the Prandtl number in a thin shell. At $Pr = 145$ (ethylene glycol), the limit was studied where the internal dissipation determines the picture. The strong viscosity practically damped the rotation influence and the convection onset was determined mainly by viscosity and heat conduction. For $E > 0.1$ the criterium of the convection onset was observed in the form of the Rayleigh-Benard given by Eq.(6.18).

For smaller values of the Prandtl number ($Pr = 5.7$), the experimental results are described better by the asymptotic theory. As in the case of wide gap the main theoretical dependencies are confirmed and again the coefficient in the dependence of the critical Rayleigh number on the Ekman number differs by a factor of two in the experiments and theory. The azimuthal wave number was observed to increase from 2 to 4 for the Ekman number increasing from 0.02 to 0.1 while for the Rayleigh-Benard convection it is constant and equals 3.117.

Unique results were obtained by Hart et al (1986) who studied the convection in a hemispherical shell on the Space Shuttle in the absence of the homogeneous gravity force field. The radial spherical symmetric gravity field was simulated by a strong externally applied electric field and as the working fluid a dielectric was used. The thermal field was spherically symmetric and homogeneous but also various temperature distributions were created in the meridional direction from the equator to the pole. Numerical modelling was also carried out and the results agreed well with what was observed.

New types of wave motions were observed. Beside the banana-like cells arising in the lower latitudes, the wave motions were observed in temperature latitudes which interact with the cells. When the temperature was changing with latitude, "spiral waves" near the pole and "triangular waves" caused by interaction of perturbations near the equator and the middle latitudes were seen. The fields were spectrally analysed and the spectral peaks of the experimentally observed and the computed fields agree well.

Zhang & Gubbins (1992,1993) considered convection in a rotating spherical fluid shell with inhomogeneous temperature at the boundary and found a number of new forms of instability. For the case of spherical geometry, they specially emphasized the non-uniform problem with two competing basic flows: baroclinic and convective. The former is associated with the boundary temperature anomalies which tend to lock the flow close to the boundary, the latter produces time-dependent convection in the form of traveling waves. The convective flow in the form of the thermal Rossby waves tries to pass over the non-uniform boundary, when for a large range of parameters the baroclinic flows can be locked into the boundary.

Chapter 8

GEOPHYSICAL AND ASTROPHYSICAL APPLICATIONS AND ANALOGIES

All natural object are rotating. The range of the periods of them allowing hydrodynamic description spreads from milliseconds (neutron stars) to many months (e.g. the planet Venus with a period of 224 days $= 1.94 \cdot 10^7$sec) differing by ten orders of magnitude, or even more. If the source of motions is internal, the convection, at least in small scales (compared to the size of the object), can be considered as free. If the main source of energy is external , as the Sun for the planetary atmospheres, the free convection can be observed only in boundary layers above the planetary surface, or above the layer where the absorption of the solar radiation per unit mass is maximal as for deep atmospheres. The global inhomogeneity of heating of a spherical planetary atmosphere by an external source like the Sun causes a global circulation of the atmosphere. Regimes and patterns of the atmospheric circulation can be classified depending on the object and its atmospheric properties and the source intensity (Golitsyn,1970,1973). Cases are possible when internal and external heat sources are comparable in their intensities (e.g. Jupiter, Saturn, Neptune). Then, a prevailing of the small scale convection or the atmospheric general circulation in their kinetic energies per unit mass depends on the height and the atmospheric composition determining the absorption of the external (solar) and internal thermal radiation.

A large part of the material presented in the preceeding chapters has as a motivation an attempt to understand and explain geophysical fluid dynamics phenomena. The main principle of presenting here the results of studies of convection in rotating fluids was an attempt to show parallel (or consequent) theoretical and experimental considerations of the problem. The final result of many such considerations was an application of the obtained results to the explanation of motions in real geophysical and astrophysical objects. Here, we shall consider the most widely observed motion regimes of these objects and their main non-dimensional parameters for assessing intensities and time scales of the motions.

8.1 Parameters and properties of some natural objects

Starting from the first systematic laboratory experiments on modeling geophysical motions, the question on correspondence of the motion in the laboratory to the ones in nature was divided into two questions: (i) external similarity of the modelled motion and the real one and (ii) the coincidence of the main non-dimensional parameters determining the main mechanisms acting within the flows in study. Unfortunately, the external similarity in appearance of the motions does not guarantee their internal similarity in the action of the mechanisms determining the flow regimes and patterns. Coincidence of all non-dimensional parameters for laboratory or theoretical studies with those of the real objects is impossible; therefore, the choice of the parameters and the interpretation of the results leave often a desire for something better. However, there is always a hope that the precise values of some non-dimensional parameters are of no importance, i.e. there is a self-similarity of some features of the regimes in study to the exact values of those parameters. Only observations can confirm or reject this hope. Here, we shall consider some approaches to the comparative analyses of non-dimensional parameters for some real and laboratory objects.

8.1.1 Plane rotating layer analogy

Many motions on geophysical and astrophysical objects can be presented as free convection of a plane rotating layer, especially for those cases when they are caused by internal energy sources, such as thermonuclear reactions in star interiors or phase separations at the mantle-liquid core interface within the Earth's body. The scale of the motions should be small compared to the object size in order to disregard conplications of sphericity. The main characteristics of such a convection have been considered in Chapter 2. The non-dimensional parameters determining such a convection are the Rayleigh, Taylor and Prandtl numbers. In the definition of all these numbers, molecular coefficients are entering. The Rayleigh number Ra introduced in section 1.3 was obtained using as the characteristic velocity scale the value $\frac{h}{k}$ where h is the layer depth and k is the thermodiffusivity and as the temperature scale the value of $\triangle T$, the temperature difference across the layer. For situations in nature, it is more pertinent to have heat fluxes at the boundaries because it is the imbalance of various forms of the energy fluxes which forces the motions in the gravity field by changing the density and generating buoyancy forces. Therefore, instead of the usual Rayleigh number, it is more natural in geophysics and astrophysics to use the Rayleigh flux number which at large values of the Nusselt number can be represented in the form using the rate of generation (dissipation) of the kinetic energy ε inside the layer per unit mass:

$$Ra_f = Ra \cdot Nu = \frac{\alpha\, g\, f^4\, h^4}{\rho\, c_p\, k^2 \nu} = \frac{\varepsilon\, h^4}{k^2\, \nu} \tag{8.1}$$

The values of quantities entering the definitions of the main non-dimensional parameters in the approximation of the plane rotating layer convection for some geophysical and astrophysical objects are presented in the Table 8.1 and Table 8.2 (after Golitsyn (1991)). Some of these values, such as the angular rotation rates, are known with some precision, others, such as molecular parameters and kinetic coefficients, are essentially functions of the temperature and, sometimes, pressure

Object	Ω, s^{-1}	$f, \frac{Wt}{m^2}$	$\frac{f}{\rho c_p}, \frac{m \cdot K}{s}$	$g, \frac{m}{s^2}$	α, K^{-1}	h, m
Earth boundary layer	10^{-4}	100	0.08	10	$3.5 \cdot 10^{-3}$	10^3
Ocean	10^{-4}	100	$2.4 \cdot 10^{-5}$	10	$2 \cdot 10^{-4}$	200
Basalt magma	10^{-4}	0.3	10^{-6}	10	$6 \cdot 10^{-5}$	$3 \cdot 10^3$
Liquid core	10^{-4}	0.1	$1.5 \cdot 10^{-7}$	10	$4 \cdot 10^{-6}$	$2 \cdot 10^6$
Jupiter atmosphere	$2.5 \cdot 10^{-4}$	10	10^{-2}	25	$6 \cdot 10^{-3}$	$2 \cdot 10^4$
Sun granulation	$4 \cdot 10^{-6}$	$6 \cdot 10^7$	$3 \cdot 10^7$	270	$1.7 \cdot 10^{-4}$	$3 \cdot 10^5$
Accretion disk	400	$3 \cdot 10^{20}$	$3 \cdot 10^{13}$	$3 \cdot 10^9$	10^{-7}	$2 \cdot 10^4$
Cold neutron star	$4 \cdot 10^3$	10^{19}	$2 \cdot 10^5$	10^{12}	10^{-7}	0.2
Hot neutron star	$4 \cdot 10^3$	10^{22}	$2 \cdot 10^9$	10^{12}	10^{-9}	10
Jupiter core	$2.5 \cdot 10^{-4}$	10	10^{-6}	30	$2 \cdot 10^{-5}$	10^7
Deep convection in ocean	10^{-4}	100	$2 \cdot 10^{-5}$	10	10^{-4}	$2 \cdot 10^3$

Table 8.1 The parameters of some geophysical and astrophysical objects

and only their characteristic values will be used. A similar situation exists also for the rotation rates for which the value of the vector $\overline{\Omega}$ is known but in considering the motions in the approximation of the plane layer, one should take the Coriolis parameter $2\Omega \sin \theta$ with θ as the latitude. For exotic objects such as neutron stars, accretion disks, and basalt magmas we shall take only the characteristic orders of the values.

We give some short comments to the parameter estimates entering Table 8.1and 8.2. Griffits (1987) was the first who paid attention in this context to cavities with magmas feeding volcanoes. The sizes of the cavities are 1 to 10 km in height and considerably larger in the horizontal direction. The magmas are of the granite and basalt type. The parameters of the latter are such that the motions within the cavities can feel the Earth's rotation. For the liquid core (the motion in which generates and supports the geomagnetic field) the data are those of Stevenson (1977) and correspond to those of molten iron. The value of the heat flux is taken after Wood (1977) from an estimate of the power needed to support the geomagnetic field against

Object	$\varepsilon, \frac{m^2}{s^3}$	$k, \frac{m^2}{s}$	$\nu, \frac{m^2}{s}$	Ra_f	Ta	Pr
Earth boundary layer	$3 \cdot 10^{-3}$	$2 \cdot 10^{-5}$	$1.5 \cdot 10^{-5}$	$5 \cdot 10^{23}$	20	0.75
Ocean	$5 \cdot 10^{-8}$	$1.4 \cdot 10^{-7}$	10^{-6}	$2 \cdot 10^{21}$	$6 \cdot 10^{13}$	7.1
Basalt magma	$6 \cdot 10^{-10}$	10^{-6}	10^{-2}	10^{19}	$4 \cdot 10^{10}$	10^{4}
Liquid core	10^{-12}	10^{-6}	10^{-7}	$2 \cdot 10^{31}$	$6 \cdot 10^{31}$	0.1
Jupiter atmosphere	$7 \cdot 10^{-4}$	$2 \cdot 10^{-4}$	$2 \cdot 10^{-4}$	$2 \cdot 10^{24}$	$2 \cdot 10^{16}$	1
Sun granulation	$1.4 \cdot 10^{6}$	1	1	10^{26}	$2 \cdot 10^{13}$	1
Accretion disk	10^{36}	0.25	$4 \cdot 10^{-3}$	$2 \cdot 10^{32}$	10^{29}	$1.6 \cdot 10^{-2}$
Cold neutron star	$2 \cdot 10^{10}$	10^{-8}	10^{-8}	$3 \cdot 10^{30}$	10^{21}	1
Hot neutron star	10^{12}	10^{-3}	10^{-3}	10^{25}	10^{18}	1
Jupiter core	10^{-9}	10^{-6}	10^{-7}	10^{38}	$3 \cdot 10^{35}$	0.1
Deep convection in ocean	$2 \cdot 10^{-8}$	$1.4 \cdot 10^{-7}$	10^{-6}	$2 \cdot 10^{25}$	$6 \cdot 10^{17}$	7.1

Table 8.2 The parameters of some geophysical and astrophysical objects (continuation of table 8.1)

the dissipation due to the ohmic losses.

For Jupiter, the atmospheric parameters are taken at the pressure $P = 1\ atm$ (the cloud upper boundary is found to be about 0.5 atm). For all objects the kinetic coefficients ν and k were estimated from the relationships ν (or k) $\approx c_s\lambda_m$ where c_s is the sound velocity and λ_m is the molecular free path. From this and according to their definitions, we have $Ra_f \propto \rho^{-4}$ and $Ta \propto \rho^{-2}$. Therefore, going down with the density increase one can find layers at which the rotation could play a larger role than within the Jupiter cloud layer. The parameter values for lower heights where the hydrogen (the main atmospheric component on Jupiter) has a phase transition to the metallic state are taken after Hubbard (1980) and Zharkov (1983). The vertical scale $h = 10^4$ km is a substantial part of the Jupiter radius $7 \cdot 10^4$ km. For the metallized hydrogen, as for any fluid in the gravity field, there is the scale $H_s = \frac{c_p}{\alpha g}$ which for Jupiter is about $2 \cdot 10^4$ km (for the water on Earth, it is equal to $2 \cdot 10^3$ km). Since for Jupiter the scale is of the order of its radius the results for its deep interior should be taken with care though errors by order of magnitudes are hardly possible. For the Earth's liquid core the scale H_s is about 10^5 km and here, in this sense, all is in order.

For cold neutron stars, which are usually pulsars with sizes of order 10 km, the smallest rotation period found is now $1.5 \cdot 10^{-3}$ sec. Atmospheres of neutron stars are consisting of degenerated helium plasma. Their hot state corresponds to so-called bursters where the reaction burning of helium atoms into carbon is taking place which raises the temperature in the burning zone by two orders of magnitude up to $10^9\ K$. Accretion disks are a mixture of protons, electrons, helium and other elements according to their cosmic abundances; therefore, the mean molecular weight is taken to be 0.6. The radiation viscosity is taken for them. More details and references for the objects can be found in the paper by Golitsyn (1991). For all astrophysical objects, there is problem of estimates of the rates of mass transport, mixing scales and periods, i.e. estimates of the characteristic convective motions are necessary.

8.1.2 Annulus analogy

As mentioned above, the laboratory experiments in annuli with horizontal temperature differences served during last forty five years as a basis for modeling many mechanisms of the large scale circulation of various planetary atmospheres. A detailed analysis and comparison of basic non-dimensional parameters of the atmospheric circulation in the troposphere, the ocean circulation and the results of the laboratory experiments can be found in the reviews by Fultz (1961), Hide (1969) and reviews of the atmospheric dynamics were done by Lorenz (1969) and Palmen & Newton (1971). Fultz (1961) has shown that the basic non-dimensional parameters of the Earth's atmospheric circulation, i.e. the Rossby number, $Ro = 0.11$, the thermal Rossby number $Ro_T = 0.04$ and the static stability parameter $S_z = 0.04$ and that the ones of the laboratory experiments coincide because in the laboratory these values can be easily adjusted by changing the temperature difference between the cylinders and the rotation rate. The values of the other non-dimensional parameters such as the Reynolds and Nusselt numbers, the geometrical parameter $\delta_h = \frac{h}{b-a}$, the ratio of the fluid depth in an annulus to the width of the gap between the cylinders differ by several orders of magnitude. Therefore, the laboratory modelling gives a good correspondence to the atmospheric process for which the values of Ro, Ro_T and S_z are determining and the other parameters do not influence the flow patterns. The phenomena considered in Chapters 3 and 4 are examples of this.

In Chapter 3, we have mentioned that the results of modelling in annuli of the flow patterns depend only weakly on the layer depth if the depth is sufficiently large for the formation of stable rising and sinking motions on the cylinder walls. Therefore, differences in the values of the geometrical parameter δ are not significant for comparing the results of laboratory experiments and atmospheric circulation patterns. At the same time, at considering the influence of heating inhomogeneities at the planetary surface or topographic effects, the relationships between the sizes and magnitudes of the inhomogeneities with the basic parameters become important and an additional geometric parameter, $\delta_H = \frac{h}{H}$, arises where H is the size of the inhomogeneity. In this case, one can choose a value of δ_H in the laboratory corresponding to the real geophysical situation. A similar situation exists for a point or spot source of convection (Chapter 5); however, here a substantial influence of viscosity is possible at the edges of the sources. This can cause the influence of the Reynolds number on the difference between the results of the laboratory experiments and of observations.

The studies of the influence of the vertical temperature gradient or the parameter of the static stability, S_z, on the motions in the annuli was stimulated by the observed change of the atmospheric circulation on Mars during global dust storms and by the possibility of the atmospheric change on Earth as a result of injection of a large quantities of smoke after large scale nuclear war, the phenomenon of so-called "nuclear winter" (see Turk et al,1983, Golitsyn & Ginsburg,1985, Turco & Golitsyn,1988, Budyko et al,1988). In both cases, large quantities of absorbing aerosols, i.e. dust and smoke, are injected into the atmosphere, the solar radiation reaches the planetary surfaces in severly reduced quantities and is absorbed within the atmosphere and heats it up. Thus the surface is cooled and the upper layers of the atmosphere are warmed up leading to large increases of the static stability parameter S_z. This can lead to changes of the circulation patterns. On Mars the normal wave regime with $Ro_T \approx 0.1$ and $G^{-1} = \left(\frac{a\Omega^2}{g}\right)^{-1} = 4.6 \cdot 10^{-3}$ with $m = 4$ is observed but during the global dust storms a symmetric regime has been detected without cyclones (see Ryan & Henry,1979). On the Earth, during the "nuclear winter" the normal wave regime with $m = 6$ or 7 may also transform into zonal symmetric regimes without cyclone waves. Boubnov & Golitsyn (1985) and Boubnov et al (1988) have proposed an explanation of these effects using a simple few mode two-layer model and confirmed its main conclusion by means of corresponding laboratory experiments with an imposed vertical temperature gradient. The change of the static parameter S_z leads to the shift of the wave regime region on the stability diagram to the right and downward. Therefore, if originally the circulation was in the wave regime, at increasing S_z it can be found in the axially (zonally) symmetric regime.

8.1.3 *Spherical layer analogy*

Results on the convection in a spherical layer have direct relevance to the large scale atmospheric dynamics on the giant planets with fast rotation as Jupiter and Saturn (see Busse,1988). On these planets, the atmospheric motions, caused not only by the horizontal inhomogeneities of the surface heating by solar radiation as in the cases of Mars and Earth but also by the internal heat flux, are also essential. The convection due to the internal heat flux was considered by Busse (1976). As a result, intensive zonal jets arise (Smith et al,1982) which are almost regular but change with periods of years. One should also pay attention to reviews by Koschmieder (1978) on the symmetric circulations in planetary atmospheres, and those by Golitsyn (1979) and Hide (1969).

8.2 Geostrophic convection

Here we consider various convective regimes at real geophysical and astrophysical objects using results obtained for convection in a plane layer (Chapter 2). They should be rather small scale phenomena for which the changes of the Coriolis parameter $2\Omega \sin\theta$ with latitude θ are not significant. The objects we shall considered are deep ocean convection (Klinger & Marshall, 1994), boundary layers on the Earth (Raasch & Etling,1991), magma chambers in the Earth mantle (Griffiths,1987), a hypothetical prehistorical magma ocean (Solomatov & Stevenson,1993), the Earth's liquid core, Solar granulation, Jupiter's deep interior, cold and hot pulsars, and accreation discs. Their properties are presented in Tables 8.1 and 8.2. In all of these rotating objects the convection is an important phenomenon. Some of these objects are represented in the regime diagram of Fig.82.

It is characteristic that there are no natural objects in the region of regular structures. In the middle between the two curves b and c are the liquid interiors of the Earth and Jupiter and accreation disks. The remaining objects are grouped near the curve c, i.e. along both sides of it. This grouping is related first of all to the large energies given off, i.e. to the heat fluxes, passing through these systems. Table 8.3 gives values of the convection characteristics for some of the natural objects.

An phenomenon as deep convection is found near the very boundary; i.e., the characteristics of deep convection can be estimated as rotating or nonrotating convection. However, the organizations of convection in the ocean in the winter seas to the bottom in the form of tubes with a diameter of several kilometers compels us to believe that the Coriolis force here plays an important role, although our simple approach gives no detailed explanations.

The planetary boundary layer under convective conditions (but not for a stable stratification!) is practically unaffected by the Earth's rotation as for the parameters determined here. However, an object such as basaltic magma, because of the large value of the Prandtl number, is probably subject to the geostrophicity condition. In the Fig.82, a dotted line indicates the relation (2.41) for $Pr = 10^4$ (see Griffiths,1987).

If the magmatic chamber is large enough, the effect of rotation on the motions in it can be important. For small chambers, the interiour motions can be estimated disregarding rotation.

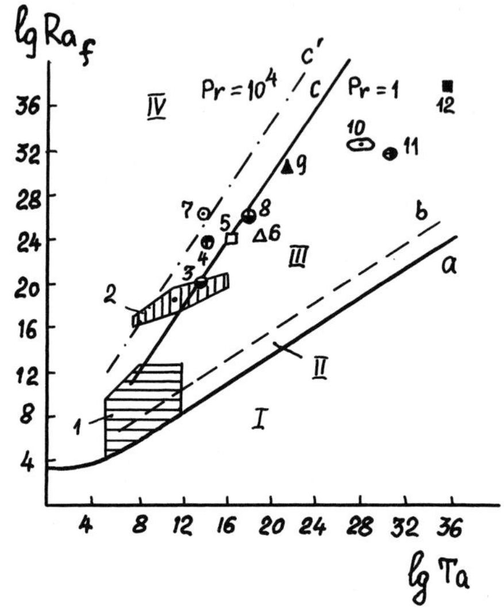

Figure 82: Convective regimes for a rotating fluid layer: I - at rest; II - region of regular motions; III - region for geostrophic irregular convection; IV - region of thermal turbulent convection; a - critical curve for onset of convection; b - boundary between regular and nonregular regimes for $Pr = 0.5$; c - boundary between geostrophic and ordinary thermal convection for $Pr = 1$; c' - same, for $Pr = 10^4$; 1 - region of parameters for the experiments; 2 - region of values for the parameters for basaltic magma; 3 - ocean, $h = 200\,m$; 4 - the atmospheric boundary layer; 5 - atmosphere of Jupiter; 6 - hot neutron star; 7 - solar granulation; 8 - ocean, $h = 2\,km$; 9 - cold pulsar; 10 - accretion disk; 11 - liquid core of the Earth; 12 - interior of Jupiter. (After Golitsyn, 1991).

We will consider in more detail the liquid core of the Earth and the interior depths of Jupiter. Here the characteristic velocities for the motions are about 5 and 100 km/yr. We note that the magnetic field intensities near the surfaces of these planets are nearly of the same ratio: about 0.5 Gauss on Earth and 10 Gauss on Jupiter. For the Earth, the velocity of 5 km/yr for the 2400 km radius is the average of the inner and outer radii for the liquid core. Such a velocity corresponds to 0.1° per year. This is close to the drift rates for the nondipole components of the geomagnetic field. We estimate the value for the magnetic Reynolds number $Re_m = \frac{ul}{\nu_m}$. For $\nu_m = 3\,m^2/s$, $l = 10^6 m = 10^3\,km$ (half the thickness of the liquid layer in the core) and $u = 1.7 \cdot 10^{-4}\,m/s$, we have $Re_m = 60$; i.e., the condition for the action of the geomagnetic dynamo, i.e. $Re_m >> 1$, is satisfied with some margin. We note that

Object	$u, m/s$	$u_T, m/s$	σ_T, K
Earth boundary layer	-	1	0.1
Ocean	$4 \cdot 10^{-2}$	$2.2 \cdot 10^{-2}$	$1.4 \cdot 10^{-2}$
Basalt magma	$4 \cdot 10^{-3}$	10^{-2}	10^{-4}
Liquid core	$1.7 \cdot 10^{-4}$	-	$1.7 \cdot 10^{-4}$
Jupiter atmosphere	2	2	$3 \cdot 10^{-3}$
Sun granulation	-	$7 \cdot 10^{3}$	500
Accretion disk	10^{-7}	$2 \cdot 10^{6}$	$2 \cdot 10^{6}$
Cold neutron star	$4 \cdot 10^{3}$	$2 \cdot 10^{3}$	200

Table 8.3 Some convective parameters for natural objects

Object	Π_Ω	Π_g	Π_M
Neutron star	80	10^{-5}	$4 \cdot 10^{-4}$
Jupiter	15	$3 \cdot 10^{-4}$	10^{-5}
Venus	$7 \cdot 10^{-3}$	10^{-3}	10^{-5}
Earth	1.43	10^{-3}	10^{-3}

Table 8.4 Similarity criteria for the general circulation of a Neutron star and some planets

$Re_m = 1$ for the motions with a scale on the order of 15 km. For Jupiter, the magnetic Reynolds number for $l = 10^7\, m$, $\nu_m = 1\, m^2/s$, and $u \sim 3 \cdot 10^{-3}\, m/s$ is equal to $3 \cdot 10^4$. Rousmaikin & Sokolov (1980) obtained an estimate for the incremental growth of the magnetic field in the dynamo-process, i.e.

$$\tau_m \sim \frac{l}{u} Re_m^{\frac{1}{3}}$$

From this, we obtain for the Earth $\tau_m \sim 10^3$ years, which is compartable to the time for large variations in the geomagnetic field. For Jupiter, the value for $\tau_m \sim 3 \cdot 10^3$ years.

Let us estimate which mass flux corresponds to a dissipation rate of $\varepsilon = 10^{-12}\, m^2/s^3$ in the liquid core. In the convection process the mass flux, because of buoyancy, is $m = \varepsilon\rho/g$. For $\rho = 10^4\, kg/m^3$, we obtain $m = 3 \cdot 10^{-2} kg/(m^2 \cdot yr)$. If this quantity is determined by differentiation at the lower boundary of the mantle, then its velocity of upward movement will be m/ρ, which in our case is $3 \cdot 10^{-6}\, m/yr$, i.e. 3 km every billion years.

For astrophysical objects we note very high rates of motion, on the order of 10 km/s for accretion disks, a few kilometers per second for cold neutron stars, and several tens of kilometers per second for hot bursters (hot neutron stars).

Large-scale motions in the atmospheres of neutron stars are the convective motions due to inhomogeneities in the heating of the atmosphere on isopotential surfaces of the gravity force. The main similarity criteria here (Golitsyn,1970,1973) are the rotational Mach number $\Pi_\Omega = \frac{\Omega r}{c}$, where r is the radius of the object and c is the speed of sound, $\Pi_g = \frac{RT}{\mu g r}$ is the ratio of the atmospheric scale height to the radius and $\Pi_M = \frac{r f}{c_p c T M}$, where T is the tempearture and M is the mass of the atmospheric column. For a cold neutron star with a smallest period of 1.5 ms, the values for these criteria are given in Table 8.4 along with the rapidly rotating Jupiter and the slowly rotating Venus (Golitsyn,1973).

We see that according to the Π_g and Π_M similarity criteria, which are independent of Ω, the pulsar is quite close to the planets in the sense that they also are much less than unity, and on the basis of them we can expect self similarity. The circulation regimes differ mainly in terms of the rotational Mach number. Our millisecond pulsar is close to Jupiter, for which there is a typical concentration of large scale motions at lower latitudes with smaller values of the Coriolis parameter. The width of this band is proportional to $\Pi_\Omega^{-\frac{1}{2}}$. If on Jupiter the width of this zone is on the order of $\pm 40°$ along both sides of the eequator, for the pulsar they should be $\left(\frac{80}{15}\right)^{\frac{1}{2}} = 2.3$ times less, and have a value of about 15°. Pulsars with a period of the order of a second have $\Pi_\Omega < 1$; i.e., from this point of view they are slowly rotating objects. For $\Pi_\Omega \sim 1$ the motions encompass all latitudes, as for the Earth.

The difference between a neutron star and a planet is that its own heat source, and not the nonuniformity in the heating of the planet by the Sun, is the moving force for the circulation. This means that in order to estimate the large-scale velocities we must take, as in the case of the Sun, for the total energy of the motions in whose atmosphere we have an estimation (Golitsyn,1972): $K = \frac{1}{2}\Omega^2 r^2 \Pi_M = \frac{1}{2} c^2 \Pi_\Omega^2 \Pi_M$. For the Sun $\Pi_\Omega = 3 \cdot 10^{-3}$. By analogy for a pulsar with $\Pi_\Omega \approx 0.1$ we obtain $U \approx c \Pi_\Omega \Pi_M^{\frac{1}{2}} = 5 \cdot 10^5 m/s \cdot 0.1 \cdot 2 \cdot 10^{-2} = 10^{-3}\, m/s = 1\, km/s$. For a radius of 10 km the nonuniformities occurring on the surface of a cold pulsar should become equalized after 10-30 s. A glance at Table 8.3 shows that the convection velocities here are half an order of magnitude greater, though they are small-scale.

BIBLIOGRAPHY

1. Al-Arabi M. & El-Ready M.K. 1976 Natural convection heat transfer from isothermal horizontal plates of different shapes. *Int. J. Heat Mass Transfer,* V.19, pp.1399-1404.
2. Al-Arabi M. & Sakr B. 1988 Natural convection heat transfer from inclined isothermal plates. *Int. J. Heat Mass Transfer,* V.31, pp.559-566.
3. Antar B.N. & Fowlis W.W. 1982 Symmetric baroclinic instability of a Hadley cell. *J.Atmos. Sci.*, V.39, pp.1280-1289.
4. Azouni A., Bolton E.W. & Busse F.H. 1986 Experimental study of convection columns in a rotating cylindrical annulus. *Geophys. Astrophys. Fluid Dyn.,* V.34, pp.301-317.
5. Barcilon V. 1964 Role of the Ekman layers in the stability of the symmetric regime obtained in rotating annulus. *J. Atmos. Sci.,* V.21, No 3, pp.291-299.
6. Barcilon V. & Drasin P.G. 1984 A weakly nonlinear theory of vacillation and baroclinic waves. *J. Atmos. Sci.*, V.41, pp.3314-3330.
7. Batchelor G.K. 1953 The conditions for dynamical similarity of the motions of a frictionless perfect-gas atmosphere. *Quart.J. Roy. Met. Soc.*, V.79, pp.224-235.
8. Behringer R.P. & Ahlers G. 1982 Heat transport and temporal evolution of fluid flow near the Rayleigh-Benard instability in cylindrical containers. *J. Fluid Mech.*, V.125, p.219.
9. Bengtsson L. & Lighthill J. (ed) 1982 *Intensive Atmospheric Vortices.* Springer-Verlag, New York, 368 p.
10. Berdnikov B.C., Grekov M.I., Malyshev V.I. & Markov V.A. 1990. Heat transfer and flow structure in heated lower layer with independent rotation of horisontal boundaries. *Proc. Inst. Heat Physics Siberian Branch Academy Sci. USSR, Novosibirsk,* V. 213, 44p.
11. Bill R.G. & Gebhart B. 1975 The transition of plane plumes. *Int. J. Heat Mass Transfer*, V.18, pp.513-526.
12. Black W.Z. & Norris J.K. 1975 The thermal structure of free convection turbulence from inclined surfaces and its influence on heat transfer. *Int. J .Heat Mass Transfer*, V.18, pp.43-50.
13. Boubnov B.M. 1977 Influence of geometry on annulus convective flow. *Izv., Atmos. Ocean Physics*, V.13, No 9, pp.621-626.
14. Boubnov B.M. 1980 An experimental study of the stability of fluid motions in a elliptical cylinder in a Coriolis force field. *Izv., Atmos. Ocean Physics,* V.16, No 4, pp.283-284.
15. Boubnov B.M. 1984a Convective fluid flows in rotating annular channel with small radius inner cylinder. *Izv., Atmos. Ocean Physics*, V.20, No6, pp.499-501.
16. Boubnov B.M. 1984b Laboratory model of convection in a rotating annular channel under conditions of horizontally and vertically inhomogeneous heating. *Izv., Atmos. Ocean Physics*, V.20, No 8, pp.679-681.

17. Boubnov B.M. 1986a Convection in a rotating nonuniformly heated annular channel with unstable stratification. *Izv., Atmos. Ocean Physics*, V.22, No 10, pp.787-791.

18. Boubnov B.M. 1986b Effect of a stable vertical temperature gradient on the motion of planetary atmospheres. *Izv., Atmos. Ocean Physics*, V.22, No 2, pp.94-101.

19. Boubnov B.M. 1987 Thermal structure of vortex convective grids. *Izv., Fluid Gas Mech.*, No 6, pp.160-166.

20. Boubnov B.M. 1988 Experimental study of thermal properties for turbulent convection in a rotating fluid layer. *Izv., Atmos. Ocean Physics*, V.24, No 9, pp.690-694.

21. Boubnov B.M. 1989 Velocity and structures of convective motions in plane rotating fluid layer. *Izv. Fluid Gas Mech.*, No. 4, pp.165-168.

22. Boubnov B.M. & Ivanov V.N. 1988 Time dependent spectrum of temperature fluctuations for free turbulent convection in a fluid layer. *Izv., Atmos. Ocean Physics*, V.24, No. 5, pp.490-498.

23. Boubnov B.M. & Gledzer E.B. 1981 On the possibility of the frequency divission in a simple convective system. *Izv. Atmos. Ocean Phys.*, V.17, pp.1254-1261.

24. Boubnov B.M. & Golitsyn G.S. 1985a Regimes convection in rotating fluids. *Dokl. AN SSSR*, V.281, No 3, pp.552-555.

25. Boubnov B.M. & Golitsyn G.S. 1985b Theoretical and experimental study of the influence of static stability on general circulation structures. *Dokl. AN SSSR*, V.281, No 5, pp.1076-1079.

26. Boubnov B.M. & Golitsyn G.S. 1988 Influence of vertical stratification on the motion patterns in differentially heated rotating annulus. *Geophys. Astrophys. Fluid Dynamics*, V.42, pp.1-35.

27. Boubnov B.M. & Golitsyn G.S. 1986 Experimental study of convective structures in rotating fluids. *J. Fluid Mech.*, V.167, pp.503-531.

28. Boubnov B.M. & Golitsyn G.S 1990a Classification and some quantitative characteristics of the convective regimes in horizontal rotating fluid layer. *Dokl. AN SSSR*, V.313, pp.55-58.

29. Boubnov B.M. & Golitsyn G.S. 1990b Temperature and velocity field regimes of convective motions in a rotating plane fluid layer. *J. Fluid Mech.*, V.219, pp.215-239.

30. Boubnov B.M. & Golitsyn G.S. 1991 Diffusion in fluid for geostrophic convective regime. *Dokl. AN SSSR*, V.317, pp.602-605.

31. Boubnov B.M. & Golitsyn G.S. 1993 Thermal turbulence and rotation. *Russ. J. Comp. Math.*, V.1, No 1, pp.59-70.

32. Boubnov B.M., Golitsyn G.S & Senatorsky A.O. 1991 Convection in a rotating cylindrical annulus with azimuthally non-uniform heating. *Geophys. Astrophys. Fluid Dynamics*, V.57, p.p.1-18.

33. Boubnov B.M., Golitsyn G.S. & Filatov V.A. 1992 Heat transfer in plane rotating fluid layer on large Rayleigh numbers. *Dokl. AN SSSR*, V.325, pp.684-686.

34. Boubnov B.M. & Heijst van G.J.F. 1993 On the frequency of temperature fluctuations in convection from a local heat source. *Izv., Atmos. Ocean Phys.*, V.29, No 2, pp.188-

193.

35. Boubnov B.M. & Heijst van G.J.F. 1994 Experiments on convection from horizontal plate with and without rotation. *Experiments in Fluids,* V.16, p.p.155-164.

36. Boubnov B.M. & Linden P.F. 1994 Diffusion circulation. Submitted to *J. Fluid Mech.*

37. Boubnov B.M. & Senatorsky A.O. 1989a Experimental study of the convection onset in plane rotating fluid layer. *Izv,. Fluid Gas Mech.*, No 5, pp.177-179.

38. Boubnov B.M. & Senatorsky A.O. 1989b Experimental study of convective flows in rotating ring-shaped vessels with zonally inhomogeneous heating. *Izv., Atmos. Ocean Physics*, V.25, No 3, pp.167-171.

39. Boussinesq J. 1903 *Theorie analytique de la chaleur, V.2*, Paris.

40. Brugge R., Jones H.L. & Marshall J.C. 1991 Non-hydrostatic ocean modelling for studies of open-ocean deep convection. In *Deep Convection and Deep Water Formation in the Oceans*, Chu P.C. & Gascard J.C. (eds.), pp.325-340, Elsevier.

41. Brunsvold A.R. & Vest C.M. 1973 The effect of shear and stratification on the stability of rotating fluid layer. *J. Fluid Mech.*, V.59, pt.2, pp.369-390.

42. Buell J.C. & Catton J. 1983 Effects of the rotation on the stability of a bounded cylindrical layer of fluid heated from below. *Phys. Fluids*, V.26, pp.892-896.

43. Busse F.H. 1970 Thermal instabilities in rapidly rotating system. *J. Fluid Mech.*, V.44, pp.441-460.

44. Busse F.H. 1976 A simple model of convection in the Jovian atmosphere. *Icarus*, V.20, pp.255-260.

45. Busse F.H. 1981. Transition to turbulence in Rayleigh-Benard convection. In *Hydrodynamic Instabilities and Transition to Turbulence.* Swinney H.L. & Gollub J.R. (eds.), Springer Verlag.

46. Busse F.H. 1982 Thermal convection in rotating systems. *Proc. Ninth US Natl. Cong. Appl. Math., Am. Soc. Mech. Engineers*, New York, pp.299-305.

47. Busse F.H. 1983 A model of mean zonal flow in the major planets. *Geophys. Astrophys. Fluid Dyn.,* V.23, pp.153-174.

48. Busse F.H. 1984 Transition to turbulence via the statistical limit cycle route. In *Turbulence and Chaotic Phenomena in Fluids* (ed. by T. Tatsumi), pp.197-202, Elsevier.

49. Busse F.H. 1986 Asymptotic theory of convection in a rotating cylindrical annulus. *J. Fluid Mech.,* V.173, pp.545-556.

50. Busse F.H. & Carrigan C.R. 1974 Convection induced by centrifugal buoyancy. *J. Fluid Mech.*, V.62, pp.579-592.

51. Busse F.H. & Clever R.M. 1979a Instabilities of the convection rolls in a fluid of moderate Prandtl number. *J. Fluid Mech.*, V.91, pt. 2, pp.465-538.

52. Busse F.H. & Clever R.M. 1979b Nonstationary convection in a rotating system. In: *Recent developments in theoretical and experimental fluid mechanics.* V.Muller, K.G. Roesher, B.Schmidt (eds.), Springer, Berlin, Heiderberg, New York, pp. 376-385.

53. Busse F.H. & Chen W.L. 1981 On the (nearly) symmetric instability. *J. Atmos. Sci.*, V.39, pp.877-880.

54. Busse F.H. & Cuong P.G. 1977 Convection in rapidly rotating spherical fluid shells. *Geophys. Astrophys. Fluid Dynamics*, V.8, No 1, pp.17-41.

55. Busse F.H. & Heikes K.E. 1980 Convection in a rotating annulus. *Science,* V.208, pp.173-175.

56. Busse F.H. & Hood L.L. 1982 Differential rotation driven by convection in a rotating annulus. *Geophys. Astrophys. Fluid Dyn.,* V.21, pp.59-74.

57. Buzina G., Pfeffer L.R. & Kung R. 1978 Cyclic variations of the imposed temperature contrast in thermally driven rotating annulus of fluid. *J. Atmos. Sci.*, V.35, No 5, pp.859-881.

58. Buzina G., Pfeffer L.R. & Kung R. 1984 Transition to geostrophic turbulence in rotating differentially heated annulus of fluid. *J. Fluid Mech.*, V.145, pp.377-404.

59. Buzina G. & Veronis G. 1971 Spin-up of a stratified fluid. *J. Fluid Mech.*, V.50, pp.579-608

60. Carrigan C.R. & Busse F.H. 1983 An experimental and theoretical investigation of the onset of convection in rotating spherical shells. *J. Fluid Mech.*, V.126, pp.287-305.

61. Chan S.K. 1974 Investigation of turbulent convection under a rotational constraint. *J. Fluid Mech.*, V.64, pt.3, pp.477-506.

62. Chandrasekhar S. 1953 The instability of a layer of fluid heated from below and subject to Coriolis forces. *Proc. Roy. Soc. London*, A 217, No1130, pp.306-327.

63. Chandrasekhar S. 1961 *Hydrodynamic and Hydromagnetic Stability.* Claredon, Oxford Univ. Press, 652 p.

64. Charlson G.S. & Sani R.L. Finite amplitude axisymmetric thermoconvective flows in a bounded cylindrical layer of fluid. *J. Fluid Mech.*, V.71, pp.209-230.

65. Charney J.G. & DeVore J.G. 1979 Multiple flow equilibria in the atmosphere and blocking. *J. Atmos. Sci.*, V.36, pp.1205-1216.

66. Charney J.G. & Straus D.M. 1980 Form-drag instability, multiple equilibria and propagating planetary waves in baroclinic orographycally forced wave system. *J. Atmos. Sci.*, V.37, pp.1157-1176.

67. Chen R., Fernando H.J.S. & Boyer D.L. 1989 Formation of isolated vortices in a rotating convecting fluid. *J. Geophys. Res.,*V.94 (D15), pp.18445-18453.

68. Church C.R., Snow J.T. & Agee E.M. 1977 Tornado vortex similations at Pardue University. *Bulletin of Amer. Meteor. Soc.*, V.58, No 9, pp.900-908.

69. Church C.R., Snow J.T., Agee E.M., Baker G.L. & Barnhart B.J. 1977 Experimental vortex configurations as a function of increasing swirl ratio. *Tenth conference of severe local storms.* Omaha, Nebraska.

70. Clever R.M. & Busse F.H. 1977 Instabilities of longitudinal convective rolls in an inclined layer. *J. Fluid Mech.*, V.81, pp.107-127.

71. Clever R.M. & Busse F.H 1978 Large wavelength convection rolls in low Prandtl number fluids. *J. Appl. Math. Phys.*, V.29, pp.711-715.

72. Clever R.M. & Busse F.H. 1979 Nonlinear properties of convection rolls in a horizontal layer about vertical axis. *J. Fluid Mech.*, V.94, pp.609-627.

73. Daniels P.G. 1980 The effect of centrifugal acceleration on axisymmetric convection in a shallow rotating cylinder or annulus. *J. Fluid Mech.*, V.99, pp.65-84.

74. Deardorff J.W. 1965 Gravitational instability between horizontal plates with shear. *Phys. Fluids*, V.8, pp.1027-1030.

75. Deardorff J.W. & Willis G.E. 1967 Investigation of turbulent thermal convection between horizontal plates. *J. Fluid Mech.*, V.28, pt.4, pp.675-705.

76. Dikarev S.N. 1983 On the influence of rotation on the convective structure in deep homogeneous fluid. *Dokl AN SSSR,* V.273, pp.718-720.

77. Douglas H.A. & Mason P.J. 1973 Thermal convection in large rotating fluid annulus: some effects of varying the aspects ratio. *J. Atmos. Sci.,* V.30, No 6, pp.1124-1134.

78. Eady E.T. 1949 Long waves and cyclone waves. *Tellus*, V.1, pp.33-52.

79. Ecke R.E., Zhong F. & Knobloch E. 1992 Hopf bifurcation with broken reflection symmetry in rotating Rayleigh-Benard convection. *Europhys. Lett.*, V.19, pp.177-182.

80. Flasar F.M. & Gierasch P.J. 1978 Turbulent convection within rapidly rotating superadiabatic fluids with horizontal temperature gradients. *Geophys. Astrophys. Fluid Dynamics*, V.10, No 3, pp.175-212.

81. Fein J.S. 1978 *Boundary layers in Homogeneous and Stratified Rotating Fluids.* Univ. Press of Florida.

82. Fein J.S. & Pfeffer R.L. 1976 An experimental study of the effects of Prandtl number on thermal convection in a rotating, differentially heated cylindrical annulus of fluid. *J.Fluid Mech.*, V.75, pt.1, pp.81-112.

83. Fernando H.J.S., Boyer D.L. & Chen R. 1989 Turbulent thermal convection in rotating and stratified fluids. *Dynamics Atmos. Oceans,* V.13, pp.95-121.

84. Fernando H.J.S., Chen R. & Boyer D.L. 1991 Effects of rotation on convective turbulence. *J. Fluid Mech.*, V.228, pp.513-547.

85. Fitzjarrald D.E. 1976 An experimental study of turbulent convection in air. *J. Fluid Mech.*, V.73, pt.4, pp.693-721.

86. Fitzjarrald D.E. 1982 An investigation of wave-amplitude vacillation using a light-speckle velocity measuring technique. *J. Phys. E:Sci. Instrum.*, V.15, pp.911-915.

87. Foster T.D. & Waller S. 1985 Experiments on convection at very high Rayleigh numbers. *Phys. Fluids,* V.28, No22, pp.455-461.

88. Fowlis W.W. & Hide R. 1965 Thermal convection in rotating annulus of liquid: effects of viscosity on the transition between axisymmetric and nonaxisymmetric flow regimes. *J. Atmos. Sci.,* V.22, No 5, pp.541-558.

89. Fujii T. & Umara R. 1972 Natural convection heat transfer from a plate with arbitrary inclination. *Int. J. Heat Mass Transfer,* V.15, pp.755-767.

90. Fultz D 1951 Experimental analogies to atmospheric motions. In: *Compedium of Meteorology (Ed. Malone T.F.),* Bostone, pp.1235-1248.

91. Fultz D.1961 Developments in controlled experiments on large scale geophysical problems. *Advances in Geophysics*, V.7, pp.1-104.

92. Fultz D., Loy R., Owens G., Boham W., Taylor R. & Weil J. 1958 Studies of thermal convection in a rotating cylinder with some implications for large-scale atmospheric motions. *Meteor. Mon.*, V.4, pp.77-101.

93. Fultz D., Nakagawa Y. & Frenzen P. 1954 An instance in thermal convection of Eddington's "overstability". *Phys. Rev.*, V.94, pp.1471-1472.

94. Fultz D., Nakagawa Y. & Frenzen P. 1955 Experiments on overstable thermal convection in mercury. *Proc. Roy. Soc. London*, A, V.231, pp.211-225.

95. Gebhard B. 1973 Instability, transition and turbulence in buoyancy induced flows. *Ann. Rev. Fluid Mech.*, V.5, pp.213-246.

96. Ginzburg A.I., Golitsyn G.S. & Fedorov K.N. 1979 Measurements of the time scale of convection in a liquid cooled from the surface. *Izv., Atmos. Ocean Physics*, V.15, No 3, pp.227-228.

97. Glatzmaier G.A. & Gilman P.A. 1981 Compressible convection in a rotating spherical shell. A linear innelastic model. *Astrophys. J. Suppl.*, V.45, pp.351-380.

98. Goldstein H.F., Knobloch E., Mercader I. & Net M. 1993 Convection in a rotating cylinder. Part 1 Linear theory for moderate Prandtl numbers. *J. Fluid Mech.,* V.218, pp.583-604.

99. Goldstein R.J., Chiang H.D. & See D.L. 1990 High Rayleigh number convection in horizontal enclosure. *J. Fluid Mech.*, V.213, pp.111-126.

100. Goldstein R.J. & Lau K.S. 1983 Laminar natural convection from a horizontal plate and the influence of plate-edge extension. *J. Fluid Mech.*, V.129, pp.55-75.

101. Golitsyn G.S. 1970 A similarity approach to the general circulation of planetary atmospheres. *Icarus,* V.13, No 1, pp.1-24.

102. Golitsyn G.S. 1972 Use of the similarity arguments for studies of the Sun's general circulation. *Astronomical Journal,* V.49, No 2, pp.350-360, (in Russian).

103. Golitsyn G.S. 1973 *Introduction to the Dynamics of the Planet Atmospheres.* Hydrometeoizdat, Leningrad, 104 p.(In Russian).

104. Golitsyn G.S. 1979 Simple theoretical and experimental study of convection with some geophysical applications and analogies. *J. Fluid Mech.*, V.95, pp.567-608.

105. Golitsyn G.S. 1980a Geostrophic convection. *Dokl. AN SSSR,* V.251, pp.1356-1360.

106. Golitsyn G.S. 1980b *Study of the Convection with Some Geophysical and Astrophysical Applications.* Hydrometeoizdat , Leningrad, 55p.(In Russian).

107. Golitsyn G.S. 1981 Structures of convection in rapid rotation. *Dokl. AN SSSR,* V.261, No 2, pp.317-320.

108. Golitsyn G.S. 1991 Convection regimes on various rotating geophysical and astrophysical objects. *Izv., Atmos. Ocean Phys.,* V.27, pp.13-21.

109. Golitsyn G.S. & Ginsburg A.S. 1985 Comparative estimates of climatic consequences of Martin dust storms and of possible nuclear war. *Tellus,* V.37B, No 3, pp.173-181.

110. Goncharov V.P. & Gryanic V.M. 1986 Dynamics of solitary dissipative vortices. Vortex lattices and their stability. *J. Exper. Theor. Physics,* V.91, No 5, pp.1653-1665.

111. Goroff I.R. 1960 An experiment on heat transfer by overstable and ordinary convec-

tion. *Proc. Roy. Soc. London,* A, V.254, pp.537-541.

112. Grachev A. A. 1989 Temperature profile and heat transfer in free turbulent convection over horizontal surface. *Izv., Atmosph.Oceanic Physics,* V.25, pp.124-131.
113. Grachev A.A. & Yaroshevich M.J.1989 Measurement of heat transfer by free convection in water cooling from the free surface. *Izv., Atmos. Ocean Phys.*, V.25, No 8, pp.887-890.
114. Green J.S.A. 1970 Transfer properties of the large-scale eddies and the general circulation of the atmosphere. *Quart. J. Roy Meteor. Soc.*, V.96, pp.157-185.
115. Greenspan H.P. 1968 *The Theory of Rotating Fluids.* Cambridge Univ. Press, 304p.
116. Griffiths R.W. 1987 Effects of the Earth's rotation on convection in magma chambers. *Earth and Planet Sci. Letts.*, V.85, pp.525-536.
117. Griffiths R.W. & Linden P.RF. 1981 The stability of vortices in a rotating, stratified fluid. *J. Fluid Mech.*, V.105, pp.283-316.
118. Griffiths R.W. & Linden P.F. 1985 Intermittent baroclinic instability and fluctuations in geophysical circulations. *Nature,* V.316, No 6031, pp.801-803.
119. Gubbins D. & Masters T.G. 1979 Driving mechanism for the Earth's dynamo. *Adv. Geophys.*, V.21, pp.51-86.
120. Guoqing L., Kung R. & Pfeffer L.R. 1987 An experimental investigation on topographyc effect on multiple regimes of baroclinic flow. *Scientica Sinica (ser.B),* V.30, No 1, pp.83-95.
121. Hadlock P.K., Na J.Y. & Stone P.H. 1972 Direct thermal verification of symmetric baroclinic instability. *J. Atmos. Sci.,* V.29, No 7, pp.1391-1393.
122. Hart J.E. 1971 Transition to a wavy vortex regime in convective flow between inclined plates. *J. Fluid Mech.*, V.48, pp.265-271.
123. Hart J.E., Glatzmaier G.A. & Toomre J. 1986 Space laboratory and numerical simulations of thermal convection in a rotating hemispherical shell with radial gravity. *J. Fluid Mech.*, V.173, pp.519-544.
124. Hartman D.L. & Shart D.A. 1979 On the role of zonal assymetry in climate change. *J. Atmos. Sci.*, V.36, No 3, pp.519-528.
125. Hathaway D.H., Gilman P.A. & Toomre J. 1979 Convective instability when the temperature gradient and rotation vector are oblique to gravity. 1. Fluids without diffusion. *Geophys. Astrophys. Fluid Dynamics*, V.13, No 4 , pp.289-316.
126. Hathaway D.H., Gilman P.A. & Toomre J. 1979 Convective instability when the temperature gradient and rotation vector are oblique to gravty. 2. Real fluids with effects of diffusion. *Geophys. Astrophys. Fluid Dynamics*, V.15, No 1-2 , pp.7-37.
127. Hathaway D.H. & Fowlis W.W. 1986 Flow regimes in a shallow rotating cylindrical annulus with temperature gradients imposed on the horizontal boundaries. *J. Fluid Mech.*, V.172, pp.401-418.
128. Hathaway H.D. & Somerville R.C.J. 1983 Three-dimensional simulations of convection in layers with tilted rotation vectors. *J.Fluid Mech.*, V.126, pp.75-89.
129. Hathaway H.D. & Somerville R.C.J. 1985 Numerical simulation in three space dimen-

sions of time-dependent thermal convection in a rotating fluid. *Lectures in Applied Mathematics*, V.22, pp.309-319.

130. Hathaway D.H. & Somerville R.C.J. 1986 Nonlinear interactions between convection, rotation and flows with vertical shear. *J. Fluid Mech.*, V.164, pp.91-105.

131. Herrmann J. & Busse F.H. 1993 Asymptotic theory of wall-attached convection in a rotating fluid layer. *J.Fluid Mech.,* V.255, pp.183-194.

132. Hide R. 1953 Some experiments on thermal convection in rotating fluids. *Quart. J. Roy. Meteor. Soc.*, V.78, No 339, p.161.

133. Hide R 1958 An experimental study of thermal convection in a rotating liquid. *Phil. Trans. Roy. Soc. London,* V. 250, No 983, pp.441-478.

134. Hide R. 1967 Theory of axisymmetric thermal convection in a rotating fluid annulus. *Phys. Fluids*, V.10, pp.56-68.

135. Hide R. 1969 Some laboratory experiments on free thermal convection in a rotating fluid subject to a horizontal temperature gradient and their relation to the theory of the global atmospheric circulation. In: *The Global Circulation of the Atmosphere,* G.A. Corby (ed), London Roy .Meteor. Soc., pp.196-221.

136. Hide R & Mason P.J. 1970 Baroclinic waves in a rotating fluid subject to internal heating. *Phil. Trans. Roy. Soc. London*, V.268, No 1186, pp.201-232.

137. Hide R. & Mason P.J. 1975 Sloping convection in a rotating liquid. *Adv. Physics*, V.24, No 1, pp.47-100.

138. Hignett P., Ibberson A. & Killworth P. 1981 On rotating thermal convection driven by non-uniform heating from below. *J. Fluid Mech.*, V. 109, pp.161-187.

139. Homsy G.H. & Hudson J.L. 1969 Centrifugally driven convection in a rotating cylinder. *J. Fluid Mech.*, V.35, pp.33-52.

140. Homsy G.H. & Hudson J.L. 1971a The asymptotic stability of a bounded rotating fluid heated from below: conductive basic state. *J. Fluid Mech.*, V.45, pp.353-373.

141. Homsy G.H. & Hudson J.L. 1971b Centrifugal convection and its effect on asymptotic stability of a bounded rotating fluid heated from below. *J. Fluid Mech.*, V.48, pp.605-624.

142. Hopfinger E.L. (Ed.) 1992 *Rotating Fluids in Geophysical and Industrial Applications* Springer Verlag, Wien-New York.

143. Hopfinger E.L., Browand F.K. & Gange Y. 1982 Turbulence and waves in rotating tank. *J. Fluid Mech.*, V.125, pp.505-534.

144. Howard L.N. 1963 Heat transport by turbulent convection. *J. Fluid Mech.*, V.17, pt 3, pp.405-432.

145. Hubbard W.B. 1980 Intrinsic luminosities of the Jovian planets. *Rev. Geopys. Space Phys.,* V.18, No 1, pp.1-10.

146. Hunter C. 1967 The axisymmetric flow in a rotating annulus due to a horizontally applied temperature gradient. *J. Fluid Mech.*, V.27, pp.753-778.

147. Hunter C. & Riahi N. 1975 Nonlinear convection in a rotating fluid. *J. Fluid Mech.*, V.72, pp.433-454.

148. Husar R.B. & Sparrow E.M. 1968 Patterns of free convection flow adjacent to horizontal heated surfaces. *Int. J. Heat Mass Transfer*, V.11, pp.1206-1208.

149. Jngersoll A.P. 1966 Convective instability in plane Couette flow. *Phys. Fluids*, V.9, pp.682-689.

150. Kaiser J.A.C. 1970 Rotating deep annulus convection. Part 2. Wave instabilities, vertical stratification and associated theories. *Tellus*, V.22, pp.275-287.

151. Katsaros K.B., Liu W.T., Businger J.A. & Tillman J.E. Heat transport and thermal structure in the interfacial boundary layer measured in an open tank of water in turbulent free convection. *J. Fluid Mech.*, V.83, pp.311-335.

152. Ketchum C.B. 1972 An experimental study of baroclinic annulus waves. *J. Atmos. Sci.*,V.29, No 4, pp.665-679.

153. Klinger B.A. & Marshall J. 1994 Regimes and scaling laws for rotating deep convection in the ocean. *Dyn. Atmos. Ocean,* to be published.

154. Kirdjashkin A.G. & Semenov B.I. 1983 Temperature spectrum in vertical layer under thermal gravitational convection. *Izv., Heat Phys. High Temp.*, V.21, No 4, pp.731-739.

155. Koschmieder E.L. 1966a Convection on uniformly heated plane. *Beitr. Physik Atmos.,* V.39, pp.1-11.

156. Koschmieder E.L. 1966b Convection on nonuniformly heated plane. *Beitr. Physik Atmos.,* V.39,pp.208-216.

157. Koschmieder E.L. 1967 On convection on a uniformly heated rotating plane. *Beitr. Physik Atmos.*, V40, pp.216-225.

158. Koschmieder E.L. 1968 Convection on non-uniformly heated, rotating plane. *J. Fluid Mech.*, V.33, pp.515-527.

159. Koschmieder E.L. 1972 Convection in rotating laterally heated annulus. *J. Fluid Mech.*, V.51, pp.637-656.

160. Koschmieder E.L. 1978 Symmetric circulations of planetary atmospheres. *Advances in Geophysics*, V.20, pp.131-181.

161. Koschmieder E.L. & Lewis E.R. 1986 Hadley circulations on nonuniformly heated rotating plate. *J. Atmos. Sci.,* V.43, pp.2514-2526.

162. Kraichnan R.H. 1962 Turbulent thermal convection at arbitrary Prandtl number. *Phys. Fluids,* V. 5, pp.1374-1389.

163. Krishnamurti R. 1970 On the transition to turbulent convection. Part 1. The transition from three-dimension flow. *J.Fluid Mech.*, V.42, pt 2, pp.295-307.

164. Krishnamurti R. 1971 On the transition to turbulent convection. In: *Proc. 8th Symp. on Naval Hydrodynamics,* Rep. ARC-179, pp.289-310, Office of Naval Research.

165. Krishnamurti R. 1973 Some further studies on the transition to turbulent convection. *J. Fluid Mech.*,V.60, pp.285-303.

166. Kuo E.Y.,& Cross M.C. 1993 Travelling-wave wall states in rotating Rayleigh-Benard convection. *Phys. Rev.,* E 47, pp.2245-2248.

167. Kuo H.L. 1956 Forced and free axially symmetric convection produced by differential

heating in a rotating fluid. *J. Met.*, V.13, pp.521-527.

168. Kuo H.L. 1966 On the dynamics of convective atmospheric vortices. *J. Atmos. Sci.*, V.23, pp.25-42.

169. Kuppers G. 1970 The stability of steady finite amplitude convection in a rotating fluid layer. *Phys. Lett.*, A 32, pp.7-8.

170. Kuppers G. & Lortz D. 1969 Transition from laminar convection to thermal turbulence in rotating fluid layer. *J. Fluid Mech.*, V.35, pp.609-620.

171. Lacher R.C., McArthur R. & Buzina G. 1977 Catastrophic changes in circulation flow patterns. *Amer. Sci.*, V65, pp.614-621.

172. Ladeinde F. & Torrance K.E. 1991 Convection in a rotating, horizontal cylinder with radial and normal gravity forces. *J. Fluid Mech.*, V.228, pp.361-385.

173. Landau L.D. & Lifshits E.M. 1989 *Fluid Mechanics.* Pergamon, New York.

174. Leaman K.D. & Schott F. 1991 Hydrographic structure of convection regime in the Gulf of Lions: Winter 1987. *J. Phys. Oceanogr.*, V.21, pp.573-596.

175. Li G.Q., Kung R. & Pfeffer L.R. 1986 An experimental study of baroclinic flows with and without two-wave bottom topography. *J. Atmos. Sci.*,V.43, pp.2585-2599.

176. Lipps F.B. 1971 Two-dimensional numerical experiments in thermal convection with vertical shear. *J.Atmos. Sci.*, V.28, pp.3-19.

177. Lipps F.B. 1976 Numerical simulation of three-dimensional Benard convection in air. *J.Fluid Mech.*, V.75, pt.1, pp.113-148.

178. Lloyd J.R. & Sparrow E.M. 1970 On the instability of natural convection flow on inclined plates. *J. Fluid Mech.*, V.42, pp.465-470.

179. Lorenz E.N. 1960 Energy and numerical weather prediction. *Tellus,* V.12, pp.364-373.

180. Lorenz E.N. 1962 Symplified dynamic equation applied to the rotating basin experiments. *J. Atmos. Sci.*,V.19, pp.39-51

181. Lorenz E.N. 1963 The mechanics of vacillation. *J. Atmos. Sci.*,V.20, pp.448-464.

182. Lorenz E.N. 1969 *Nature and Theory of the General Circulation of the Atmosphere.* World Meteor. Organ., Geneva, 259 p.

183. Lucas P.G., Phontenhauer J.M. & Donnely R.J. 1984 Stability and heat transfer of rotating cryogens. Part 1. Influence of rotation on the onset of convection in ^4He. *J.Fluid Mech.*, V.129, pp.251-264.

184. Maxworthy T. & Narimosa S. 1991 Vortex generation by convection in a rotating fluid. *Ocean Modelling,* V.92, pp.1037-1040.

185. Maxworthy T. 1992 Convection and shear flow turbulence with rotation. In: *Rotating Fluids in Geophysical and Industrial Applications,* (Hopfinger E.L., ed.), Springer-Verlag, Wien-New York, pp.339-357.

186. McIntyre M.E. 1970 Diffusive destabilization of the baroclinic circular vortex. *Geophys. Fluid Dynamics,* V.1, pp.19-57.

187. McKenzie D.P., Roberts J.M. & Weiss N.O. 1974 Convection in the Earth's mantle: towards a numerical simulation. *J. Fluid Mech.*, V.62, pt. 3, pp.465-538.

188. Miller T.L. & Fowlis W.W. 1986 Laboratory experiments in baroclinic annulus with heating and cooling on the horizontal boundaries. *Geophys. Astrophys. Fluid Dynamics*, V.34, pp.283-300.

189. Monin A.S & Yaglom A.M. 1965 *Statistical Hydromechanics. Part 1* . Moscow: Nauka, 639 p. (English transl.: 1971, MIT).

190. Mullen J.B. & Maxworthy T. 1977 A laboratory model of dust-devil vortices. *Dyn. Atmos. Oceans,* V.1, pp.181-214.

191. Muller U. 1965 Undersuchungen an Rotation symmetrishen Zellularkonvektionsstrohmungen. *Beitr. Physik Atmos.*, V.38, pp.1-8.

192. Nakagawa Y. & Frenzen P. 1955 A theoretical and experimental study of cellular convection in rotating fluids. *Tellus*, V.7, pp.1-21.

193. Nieler P.P. & Bisshopp F.E. 1965 On the influence of Coriolis force on onset of the thermal convection.

194. Obukhov A.M. 1960 On the structure of the temperature field and the velocity field during free convection. *Izv. ser.Geophys.,* No 9, pp.1392-1396.

195. Or A.C. 1990 New phenomena in the Eckhaus instability of thermal Rossby waves. *J.Fluid Mech.,* V.216, pp.613-628.

196. Pedlosky J. 1965 On the stability of the baroclinic flows as a functional of the velocity profile. *J. Atmos. Sci.,* V.22, pp.137-145.

197. Pedlosky J. 1977 A model of wave amplitude vacillation. *J. Atmos. Sci.,* V.34, pp.1898-1912.

198. Pedlosky J. 1979 *Geophysical Fluid Dynamics.* Springer-Verlag, 624p.

199. Pfeffer R.L. & Barcilon A. 1978 Determination of the eddy fluxes of heat and eddy temperature variances using weakly nonlinear theory. *J. Atmos.Sci.*, V.35, No 11, pp.2099-2110.

200. Pfeffer R.L., Buzyna G. & Fowlis W.W. 1974 Synoptic features and energetics of wave-amplitude vacillation in a rotating differentially heated fluid. *J. Atmos. Sci.,* V.31, pp.622-645.

201. Pfeffer R.L., Buzyna G. & Kung R. 1980a Time dependent modes of behavior of thermally driven rotating fluids. *J. Atmos. Sci.*,V.37, No 10, pp.2129-2149.

202. Pfeffer R.L., Buzyna G. & Kung R. 1980b Relationships among eddy fluxes of heat, eddy temperature variances and basic-state temperature parameters in thermally-driven rotating fluids. *J. Atmos. Sci.,* V.37, pp.2577-2599.

203. Phontenhauer J.M., Lucas P.G. & Donnely R.J. 1984 Stability and heat trasfer of cryogens. Part 2. Effects of rotation on the heat transfer properties of convection in liquid ^4He. *J. Fluid Mech.,* V.145, pp.239-252.

204. Phontenhauer J.M., Niemela J.J. & Donnely R.J. 1987 Stability and heat transfer of rotating cryogens. Part 3. Effects of finite cylindrical geometry and rotation on the onset of convection. *J. Fluid Mech.*, V.175, pp.85-96.

205. Raasch S. & Etling D. 1991 Numerical simulation of rotating turbulent thermal convection. *Beitr. Phys. Atmosph.,* V.64, No3, pp.185-199.

206. Rayleigh 1916 On convection currents in a horizontal layer of fluid, when the higher temperature is on the under side. *Phyl. Mag.*, V.32, p.529.

207. Read P.L. 1992 Dynamics and instabilities of Ekman and Stewardson layers. In: *Rotating Fluids in Geophysical and Industrial Applications* (Hopfinger E.L., ed.), Springer Verlag, Wien-NewYork, pp.50-84.

208. Roberts P.H. 1968 On the thermal instability of a rotating fluid sphere containing heat sources. *Phil. Trans. Roy. Soc. London*, A 263, No 1136, pp.93–117.

209. Robinson A.R. 1959 The symmetric state of a rotating fluid differentially heated in the horizontal. *J. Fluid Mech.*, V.6, pp.599-620.

210. Rossby H.T. 1965 On thermal convection driven by non-uniform heating from below. An experimental study. *Deep Sea Res.*, V.12, pp.9-16.

211. Rossby H.T. 1969 A study of Benard convection with and without rotation. *J. Fluid Mech.*, V.36, pt 2, pp.309-337.

212. Rouzmaikin A.A. & Sokolov D.D. 1980 Helicity, linkage and dynamic action. *Geoph. Astroph. Fluid Dyn.*, V.16, Nos 1-2, pp.73-82.

213. Riahi N. 1977 Upper-bound problem for a rotating system. *J. Fluid Mech.*, V.81, pp.523-526.

214. Ryan J.A. & Henry R.M. 1979 Mars atmospheric phenomena during major storms, as measured at surface. *J. Geophys. Res.*, V.84, No 6, pp.2821-2829.

215. Schlichting H. 1951 *Grenzschicht-Theory.* Verlag G. Braun, Karlsruhe.

216. Schnaubelt M. & Busse F.H. 1992 Convection in a rotating cylindrical annulus Part 3. Vacillating and spatially modulated flows. *J. Fluid Mech.*,V.245, pp.155-173.

217. Snyder H.A. & Youtz E.M. 1969 Transient response of a differentially heated rotating annulus *J. Atmos. Sci.*,V.26, pp.96-99.

218. Solberg H. 1936 Le Mouvement d'inertie de l'atmosphere stable et son role dans le theorie des cyclones. In: *Proc. Union Geodesique et Geophysique Internationale Vieme Assemblee,* Edinburg, V.11, pp.66-82.

219. Solomatov V.S. & Stevenson D.J. 1993 Suspension in convective layers and style of differentiation of a terrestrial magma ocean. *J. Geophys. Res.*, V.98, No E3, pp.5375-5390.

220. Sommerville R.C.J. 1971 Benard convection in a rotating fluid. *Geophys. Fluid Dynamics,* V.2, pp.247-262.

221. Sommerville R.C.J. & Lipps F.B. 1973 A numerical study in three-dimensions of Benard convection in rotating fluid. *J. Atmos. Sci.*, V.30, No 4, pp.590-596.

222. Sparrow E.M., Goldstein G.J. & Jonsson V.K. 1964 Thermal instability in a horizontal fluid layer: effect of boundary conditions and non-linear temperature profile. *J. Fluid Mech.*, V.18, pt .4, pp.513-528.

223. Sparrow E.M. & Husar R.B. 1969 Longitudinal vorticies in natural convection from inclined surfaces. *J. Fluid Mech.*, V.37, pp.251-255.

224. Sparrow E.M., Husar R.B & Goldstein R.J. 1970 Observations and other characteristics of thermals. *J. Fluid Mech.*, V.41, pp.793-800.

225. Stewartson K. 1957 On almost rigid rotations. *J. Fluid Mech.*, V.3, pp.17-26.

226. Stone P.H. 1966 On non-geostrophic baroclinic stability. *J. Atmos. Sci.*, V.23, pp.390-400.

227. Stone P.H. 1970 On non-geostrophic baroclinic stability. Part 2. *J. Atmos. Sci.*, V.27, pp.721-732.

228. Stone P.H. 1971 Baroclinic stability under non-hydrostatic conditions. *J. Fluid Mech.*, V.45, pt.4, pp.659-671.

229. Stone P.S. 1972 A simplified radiative-dynamical model for static stability of rotating atmospheres. *J. Atmos. Sci.*, V.29, No 3, pp.405-418.

230. Stone P.H., Hess S., Hadlock P. & Ray P. 1969 Preliminary results of experiments with symmetric baroclinic instabilities. *J. Atmos. Sci.*,V.26, No 5, pp.991-996.

231. Tanaka H. & Miyata H. 1980 Turbulent natural convection in a horizontal water layer heated from below. *Int. J. Heat Mass Transfer*, V.23, No 9, pp.1273-1283.

232. Tanaki K. & Ukaji K. 1986 Stationary baroclinic eddies produced in rotating fluid annulus. *J. Meteor. Soc. Japan*, V.64, No 5, pp.681-691.

233. Thomson J. 1892 On the grand currents of the atmospheric circulation. *Phyl. Trans. Roy. Soc. London*, A183, pp.653-684.

234. Thomas D.B. & Townsend A.A. 1957 Turbulent convection over a heated horizontal surface. *J. Fluid Mech.*, V.2, pt. 5, pp.473-492.

235. Torrance K.E. 1979 Natural convection in thermally stratified encloses with localized heating from below. *J. Fluid Mech.*, V.95, pp.477-495.

236. Torrance K.E. & Rockett J.A. 1969 Numerical study of natural convection in enclosures with localized heating from below - creeping flow to the onset of laminar instability. *J. Fluid Mech.*, V.36, pp.33-54.

237. Torrance K.E., Orloff L. & Rockett J.A.1969 Experiments on natural convection in enclosures with localized heating from below. *J. Fluid Mech.*, V.36, pp.21-31.

238. Torrest M.A. & Hudson J.L. 1974 The effects of the centrifugal convection on the stability of a rotating fluid heated from below. *Appl. Sci. Res.*, V.29, No.4, pp.273-289.

239. Turco R., Toon B., Ackerman T., Pollack J. & Sagan C. 1983 Nuclear winter: Global consequences of multiple nuclear explosions. *Science*, V.222, pp.1283-1293.

240. Turner J.S. 1973 *Buoyancy Effects in Fluids.* Cambridge University Press, New York, 367p.

241. Turner J.S. 1986 Turbulent entrainment: The development of the entrainment assumption and its application to geophysical flows. *J. Fluid Mech.*, V.173, pp.431-471.

242. Veronis G.1966 Motions at subcritical values of the Rayleigh number in a rotating fluid. *J. Fluid Mech.*, V.24, pt.3, pp.545-554.

243. Veronis G. 1959 Cellular convection with finite amplitude in rotating fluid. *J. Fluid Mech.*, V.5, pp.401-435.

244. Veronis G. 1966 Motions at subcritical values of the Rayleigh number in rotating fluid. *J. Fluid Mech.*, V.24, pp.545-554.

245. Veronis G. 1968 Large-amplitude Benard convection in rotating fluid. *J. Fluid Mech.*, V.31, pp.113-140.

246. Veronis G. 1970 The analogy between rotating and stratified fluids. *Annual Review of Fluid Mechanics*, V.2, pp.37-67.

247. Vettin F. 1857 Uber den ausfsteigenden Luftstrom, die Entstehung des Hagels und der Wirbel-Sturme. *Ann. Physik*, V.102, pp.246-255.

248. Vladimirov V.V. 1985 Example of the similarity of the effects of density stratification and rotation. *Dokl. AN SSSR*, V.284, No 2, pp.310-313.

249. Von Tippelskirch H. 1959 Weitere Konvektionsversuche: Der Nachweis der Ringzellen und ihrer Verallgemeinerung. *Beitr. Physik Atmos.*, V.32, pp.2-22.

250. Williams G.P. 1967 Thermal convection in rotating fluid annulus. Part 1 and Part 2. *J. Atmos. Sci.*, V.24, pp.144-161 and pp.162-174.

251. Williams G.P. 1971 Baroclinic annulus waves. *J. Fluid Mech.*, V.49, pp.417-449.

252. Willis G.E., Deardorff J.W. & Somerville R.C. 1972 Roll-diameter dependence in Rayleigh convection and its effect upon heat flux. *J. Fluid Mech.*, V.54, pp.351-367.

253. Wood W.W. 1977 A note on the westward drift of the Earth's magnetic field. *J. Fluid Mech.*, V.82, pt.2 , pp.389-400.

254. Yavorskaya I.M. & Belyaev Yu.N. 1982 Convective motions in rotating fluid layers. *Adv. Sci. Tech. VINITY, ser. Mech. Fluid Gas.*, V.17, pp. 1-85.

255. Zhang K. & Gubbins D. 1993 Convection in a rotating spherical fluid shell with an inhomogeneous temperature boundary condition at infinite Prandtl number. *J. Fluid Mech.*, V.255, pp.209-232.

256. Zhang K.1991 Vacillatory convection in a rotating spherical fluid shell at infinite Prandtl numbers. *J. Fluid Mech.*, V.228, pp.607-628.

257. Zhang K. 1992 Spiralling columnar convection in rapidly rotating spherical fluid shells. *J. Fluid Mech.*, V.236, pp.535-556.

258. Zharkov V.N. 1983 *Internal Structure of the Earth and Planets.* Second Ed., Moscow, Fizmatgiz.

259. Zhong F., Ecke R.E. & Steinberg V. 1991 Asymmetric modes and transition to vortex structures in rotating Rayleigh-Benard convection. *Phys. Rev. Lett.*, V.67, pp.2473-2476.

260. Zierep R. 1958 Eine rotations symmetrische Zellularkonvektionsstromung. *Beitr. Phyzik Atmos.*, V.30, pp.215-222.

261. Zimin V.D. & Ketov A.I. 1978 Turbulent convection in cubic caveture heated above. *Izv., Mech. Fluid Gas,* No 1, pp.128-137.

Mechanics

FLUID MECHANICS AND ITS APPLICATIONS

Series Editor: R. Moreau

Aims and Scope of the Series

The purpose of this series is to focus on subjects in which fluid mechanics plays a fundamental role. As well as the more traditional applications of aeronautics, hydraulics, heat and mass transfer etc., books will be published dealing with topics which are currently in a state of rapid development, such as turbulence, suspensions and multiphase fluids, super and hypersonic flows and numerical modelling techniques. It is a widely held view that it is the interdisciplinary subjects that will receive intense scientific attention, bringing them to the forefront of technological advancement. Fluids have the ability to transport matter and its properties as well as transmit force, therefore fluid mechanics is a subject that is particularly open to cross fertilisation with other sciences and disciplines of engineering. The subject of fluid mechanics will be highly relevant in domains such as chemical, metallurgical, biological and ecological engineering. This series is particularly open to such new multidisciplinary domains.

1. M. Lesieur: *Turbulence in Fluids.* 2nd rev. ed., 1990 ISBN 0-7923-0645-7
2. O. Métais and M. Lesieur (eds.): *Turbulence and Coherent Structures.* 1991 ISBN 0-7923-0646-5
3. R. Moreau: *Magnetohydrodynamics.* 1990 ISBN 0-7923-0937-5
4. E. Coustols (ed.): *Turbulence Control by Passive Means.* 1990 ISBN 0-7923-1020-9
5. A.A. Borissov (ed.): *Dynamic Structure of Detonation in Gaseous and Dispersed Media.* 1991 ISBN 0-7923-1340-2
6. K.-S. Choi (ed.): *Recent Developments in Turbulence Management.* 1991 ISBN 0-7923-1477-8
7. E.P. Evans and B. Coulbeck (eds.): *Pipeline Systems.* 1992 ISBN 0-7923-1668-1
8. B. Nau (ed.): *Fluid Sealing.* 1992 ISBN 0-7923-1669-X
9. T.K.S. Murthy (ed.): *Computational Methods in Hypersonic Aerodynamics.* 1992 ISBN 0-7923-1673-8
10. R. King (ed.): *Fluid Mechanics of Mixing.* Modelling, Operations and Experimental Techniques. 1992 ISBN 0-7923-1720-3
11. Z. Han and X. Yin: *Shock Dynamics.* 1993 ISBN 0-7923-1746-7
12. L. Svarovsky and M.T. Thew (eds.): *Hydroclones.* Analysis and Applications. 1992 ISBN 0-7923-1876-5
13. A. Lichtarowicz (ed.): *Jet Cutting Technology.* 1992 ISBN 0-7923-1979-6
14. F.T.M. Nieuwstadt (ed.): *Flow Visualization and Image Analysis.* 1993 ISBN 0-7923-1994-X
15. A.J. Saul (ed.): *Floods and Flood Management.* 1992 ISBN 0-7923-2078-6
16. D.E. Ashpis, T.B. Gatski and R. Hirsh (eds.): *Instabilities and Turbulence in Engineering Flows.* 1993 ISBN 0-7923-2161-8
17. R.S. Azad: *The Atmospheric Boundary Layer for Engineers.* 1993 ISBN 0-7923-2187-1
18. F.T.M. Nieuwstadt (ed.): *Advances in Turbulence IV.* 1993 ISBN 0-7923-2282-7
19. K.K. Prasad (ed.): *Further Developments in Turbulence Management.* 1993 ISBN 0-7923-2291-6
20. Y.A. Tatarchenko: *Shaped Crystal Growth.* 1993 ISBN 0-7923-2419-6

Kluwer Academic Publishers – Dordrecht / Boston / London

Mechanics

FLUID MECHANICS AND ITS APPLICATIONS

Series Editor: R. Moreau

21. J.P. Bonnet and M.N. Glauser (eds.): *Eddy Structure Identification in Free Turbulent Shear Flows.* 1993 ISBN 0-7923-2449-8
22. R.S. Srivastava: *Interaction of Shock Waves.* 1994 ISBN 0-7923-2920-1
23. J.R. Blake, J.M. Boulton-Stone and N.H. Thomas (eds.): *Bubble Dynamics and Interface Phenomena.* 1994 ISBN 0-7923-3008-0
24. R. Benzi (ed.): *Advances in Turbulence V.* 1995 (forthcoming) ISBN 0-7923-3032-3
25. B.I. Rabinovich, V.G. Lebedev and A.I. Mytarev: *Vortex Processes and Solid Body Dynamics.* The Dynamic Problems of Spacecrafts and Magnetic Levitation Systems. 1994 ISBN 0-7923-3092-7
26. P.R. Voke, L. Kleiser and J.-P. Chollet (eds.): *Direct and Large-Eddy Simulation I.* Selected papers from the First ERCOFTAC Workshop on Direct and Large-Eddy Simulation. 1994 ISBN 0-7923-3106-0
27. J.A. Sparenberg: *Hydrodynamic Propulsion and its Optimization.* Analytic Theory. 1995 ISBN 0-7923-3201-6
28. J.F. Dijksman and G.D.C. Kuiken (eds.): *IUTAM Symposium on Numerical Simulation of Non-Isothermal Flow of Viscoelastic Liquids.* Proceedings of an IUTAM Symposium held in Kerkrade, The Netherlands. 1995 ISBN 0-7923-3262-8
29. B.M. Boubnov and G.S. Golitsyn: *Convection in Rotating Fluids.* 1995 ISBN 0-7923-3371-3
30. S.I. Green (ed.): *Fluid Vortices.* 1995 ISBN 0-7923-3376-4

Kluwer Academic Publishers – Dordrecht / Boston / London

Mechanics

SOLID MECHANICS AND ITS APPLICATIONS

Series Editor: G.M.L. Gladwell

Aims and Scope of the Series

The fundamental questions arising in mechanics are: *Why?*, *How?*, and *How much?* The aim of this series is to provide lucid accounts written by authoritative researchers giving vision and insight in answering these questions on the subject of mechanics as it relates to solids. The scope of the series covers the entire spectrum of solid mechanics. Thus it includes the foundation of mechanics; variational formulations; computational mechanics; statics, kinematics and dynamics of rigid and elastic bodies; vibrations of solids and structures; dynamical systems and chaos; the theories of elasticity, plasticity and viscoelasticity; composite materials; rods, beams, shells and membranes; structural control and stability; soils, rocks and geomechanics; fracture; tribology; experimental mechanics; biomechanics and machine design.

1. R.T. Haftka, Z. Gürdal and M.P. Kamat: *Elements of Structural Optimization.* 2nd rev.ed., 1990 ISBN 0-7923-0608-2
2. J.J. Kalker: *Three-Dimensional Elastic Bodies in Rolling Contact.* 1990 ISBN 0-7923-0712-7
3. P. Karasudhi: *Foundations of Solid Mechanics.* 1991 ISBN 0-7923-0772-0
4. *Not published*
5. *Not published.*
6. J.F. Doyle: *Static and Dynamic Analysis of Structures.* With an Emphasis on Mechanics and Computer Matrix Methods. 1991 ISBN 0-7923-1124-8; Pb 0-7923-1208-2
7. O.O. Ochoa and J.N. Reddy: *Finite Element Analysis of Composite Laminates.* ISBN 0-7923-1125-6
8. M.H. Aliabadi and D.P. Rooke: *Numerical Fracture Mechanics.* ISBN 0-7923-1175-2
9. J. Angeles and C.S. López-Cajún: *Optimization of Cam Mechanisms.* 1991 ISBN 0-7923-1355-0
10. D.E. Grierson, A. Franchi and P. Riva: *Progress in Structural Engineering.* 1991 ISBN 0-7923-1396-8
11. R.T. Haftka and Z. Gürdal: *Elements of Structural Optimization.* 3rd rev. and exp. ed. 1992 ISBN 0-7923-1504-9; Pb 0-7923-1505-7
12. J.R. Barber: *Elasticity.* 1992 ISBN 0-7923-1609-6; Pb 0-7923-1610-X
13. H.S. Tzou and G.L. Anderson (eds.): *Intelligent Structural Systems.* 1992 ISBN 0-7923-1920-6
14. E.E. Gdoutos: *Fracture Mechanics.* An Introduction. 1993 ISBN 0-7923-1932-X
15. J.P. Ward: *Solid Mechanics.* An Introduction. 1992 ISBN 0-7923-1949-4
16. M. Farshad: *Design and Analysis of Shell Structures.* 1992 ISBN 0-7923-1950-8
17. H.S. Tzou and T. Fukuda (eds.): *Precision Sensors, Actuators and Systems.* 1992 ISBN 0-7923-2015-8
18. J.R. Vinson: *The Behavior of Shells Composed of Isotropic and Composite Materials.* 1993 ISBN 0-7923-2113-8

Kluwer Academic Publishers – Dordrecht / Boston / London

Mechanics

SOLID MECHANICS AND ITS APPLICATIONS

Series Editor: G.M.L. Gladwell

19. H.S. Tzou: *Piezoelectric Shells.* Distributed Sensing and Control of Continua. 1993 ISBN 0-7923-2186-3
20. W. Schiehlen: *Advanced Multibody System Dynamics.* Simulation and Software Tools. 1993 ISBN 0-7923-2192-8
21. C.-W. Lee: *Vibration Analysis of Rotors.* 1993 ISBN 0-7923-2300-9
22. D.R. Smith: *An Introduction to Continuum Mechanics.* 1993 ISBN 0-7923-2454-4
23. G.M.L. Gladwell: *Inverse Problems in Scattering.* An Introduction. 1993 ISBN 0-7923-2478-1
24. G. Prathap: *The Finite Element Method in Structural Mechanics.* 1993 ISBN 0-7923-2492-7
25. J. Herskovits (ed.): *Structural Optimization '93.* 1995 (forthcoming) ISBN 0-7923-2510-9
26. M.A. González-Palacios and J. Angeles: *Cam Synthesis.* 1993 ISBN 0-7923-2536-2
27. W.S. Hall: *The Boundary Element Method.* 1993 ISBN 0-7923-2580-X
28. J. Angeles, G. Hommel and P. Kovács (eds.): *Computational Kinematics.* 1993 ISBN 0-7923-2585-0
29. A. Curnier: *Computational Methods in Solid Mechanics.* 1994 ISBN 0-7923-2761-6
30. D.A. Hills and D. Nowell: *Mechanics of Fretting Fatigue.* 1994 ISBN 0-7923-2866-3
31. B. Tabarrok and F.P.J. Rimrott: *Variational Methods and Complementary Formulations in Dynamics.* 1994 ISBN 0-7923-2923-6
32. E.H. Dowell, E.F. Crawley, H.C. Curtiss Jr., D.A. Peters, R. H. Scanlan and F. Sisto (eds.): *A Modern Course in Aeroelasticity.* Third Revised and Enlarged Edition. (forthcoming) ISBN 0-7923-2788-8; Pb: 0-7923-2789-6
33. A. Preumont: *Random Vibration and Spectral Analysis.* 1994 ISBN 0-7923-3036-6
34. J.N. Reddy (ed.): *Mechanics of Composite Materials.* Selected works of Nicholas J. Pagano. 1994 ISBN 0-7923-3041-2
35. A.P.S. Selvadurai (ed.): *Mechanics of Poroelastic Media.* 1995 (forthcoming) ISBN 0-7923-3329-2
36. Z. Mróz, D. Weichert, S. Dorosz (eds.): *Inelastic Behaviour of Structures under Variable Loads.* 1995 ISBN 0-7923-3397-7

Kluwer Academic Publishers – Dordrecht / Boston / London

Mechanics

From 1990, books on the subject of *mechanics* will be published under two series:
***FLUID* MECHANICS AND ITS APPLICATIONS**
Series Editor: R.J. Moreau
***SOLID* MECHANICS AND ITS APPLICATIONS**
Series Editor: G.M.L. Gladwell

Prior to 1990, the books listed below were published in the respective series indicated below.

MECHANICS: DYNAMICAL SYSTEMS
Editors: L. Meirovitch and G.Æ. Oravas

1. E.H. Dowell: *Aeroelasticity of Plates and Shells.* 1975 ISBN 90-286-0404-9
2. D.G.B. Edelen: *Lagrangian Mechanics of Nonconservative Nonholonomic Systems.* 1977 ISBN 90-286-0077-9
3. J.L. Junkins: *An Introduction to Optimal Estimation of Dynamical Systems.* 1978 ISBN 90-286-0067-1
4. E.H. Dowell (ed.), H.C. Curtiss Jr., R.H. Scanlan and F. Sisto: *A Modern Course in Aeroelasticity.* *Revised and enlarged edition see under Volume 11*
5. L. Meirovitch: *Computational Methods in Structural Dynamics.* 1980 ISBN 90-286-0580-0
6. B. Skalmierski and A. Tylikowski: *Stochastic Processes in Dynamics.* Revised and enlarged translation. 1982 ISBN 90-247-2686-7
7. P.C. Müller and W.O. Schiehlen: *Linear Vibrations.* A Theoretical Treatment of Multi-degree-of-freedom Vibrating Systems. 1985 ISBN 90-247-2983-1
8. Gh. Buzdugan, E. Mihăilescu and M. Radeş: *Vibration Measurement.* 1986 ISBN 90-247-3111-9
9. G.M.L. Gladwell: *Inverse Problems in Vibration.* 1987 ISBN 90-247-3408-8
10. G.I. Schuëller and M. Shinozuka: *Stochastic Methods in Structural Dynamics.* 1987 ISBN 90-247-3611-0
11. E.H. Dowell (ed.), H.C. Curtiss Jr., R.H. Scanlan and F. Sisto: *A Modern Course in Aeroelasticity.* Second revised and enlarged edition (of Volume 4). 1989 ISBN Hb 0-7923-0062-9; Pb 0-7923-0185-4
12. W. Szemplińska-Stupnicka: *The Behavior of Nonlinear Vibrating Systems.* Volume I: Fundamental Concepts and Methods: Applications to Single-Degree-of-Freedom Systems. 1990 ISBN 0-7923-0368-7
13. W. Szemplińska-Stupnicka: *The Behavior of Nonlinear Vibrating Systems.* Volume II: Advanced Concepts and Applications to Multi-Degree-of-Freedom Systems. 1990 ISBN 0-7923-0369-5
    Set ISBN (Vols. 12–13) 0-7923-0370-9

MECHANICS OF STRUCTURAL SYSTEMS
Editors: J.S. Przemieniecki and G.Æ. Oravas

1. L. Frýba: *Vibration of Solids and Structures under Moving Loads.* 1970 ISBN 90-01-32420-2
2. K. Marguerre and K. Wölfel: *Mechanics of Vibration.* 1979 ISBN 90-286-0086-8

Mechanics

3. E.B. Magrab: *Vibrations of Elastic Structural Members.* 1979 ISBN 90-286-0207-0
4. R.T. Haftka and M.P. Kamat: *Elements of Structural Optimization.* 1985
   Revised and enlarged edition see under Solid Mechanics and Its Applications, Volume 1
5. J.R. Vinson and R.L. Sierakowski: *The Behavior of Structures Composed of Composite Materials.* 1986 ISBN Hb 90-247-3125-9; Pb 90-247-3578-5
6. B.E. Gatewood: *Virtual Principles in Aircraft Structures.* Volume 1: Analysis. 1989
   ISBN 90-247-3754-0
7. B.E. Gatewood: *Virtual Principles in Aircraft Structures.* Volume 2: Design, Plates, Finite Elements. 1989 ISBN 90-247-3755-9
   Set (Gatewood 1 + 2) ISBN 90-247-3753-2

MECHANICS OF ELASTIC AND INELASTIC SOLIDS
Editors: S. Nemat-Nasser and G.Æ. Oravas

---

1. G.M.L. Gladwell: *Contact Problems in the Classical Theory of Elasticity.* 1980
   ISBN Hb 90-286-0440-5; Pb 90-286-0760-9
2. G. Wempner: *Mechanics of Solids with Applications to Thin Bodies.* 1981
   ISBN 90-286-0880-X
3. T. Mura: *Micromechanics of Defects in Solids.* 2nd revised edition, 1987
   ISBN 90-247-3343-X
4. R.G. Payton: *Elastic Wave Propagation in Transversely Isotropic Media.* 1983
   ISBN 90-247-2843-6
5. S. Nemat-Nasser, H. Abé and S. Hirakawa (eds.): *Hydraulic Fracturing and Geothermal Energy.* 1983 ISBN 90-247-2855-X
6. S. Nemat-Nasser, R.J. Asaro and G.A. Hegemier (eds.): *Theoretical Foundation for Large-scale Computations of Nonlinear Material Behavior.* 1984 ISBN 90-247-3092-9
7. N. Cristescu: *Rock Rheology.* 1988 ISBN 90-247-3660-9
8. G.I.N. Rozvany: *Structural Design via Optimality Criteria.* The Prager Approach to Structural Optimization. 1989 ISBN 90-247-3613-7

MECHANICS OF SURFACE STRUCTURES
Editors: W.A. Nash and G.Æ. Oravas

---

1. P. Seide: *Small Elastic Deformations of Thin Shells.* 1975 ISBN 90-286-0064-7
2. V. Panc: *Theories of Elastic Plates.* 1975 ISBN 90-286-0104-X
3. J.L. Nowinski: *Theory of Thermoelasticity with Applications.* 1978
   ISBN 90-286-0457-X
4. S. Łukasiewicz: *Local Loads in Plates and Shells.* 1979 ISBN 90-286-0047-7
5. C. Fiřt: *Statics, Formfinding and Dynamics of Air-supported Membrane Structures.* 1983 ISBN 90-247-2672-7
6. Y. Kai-yuan (ed.): *Progress in Applied Mechanics.* The Chien Wei-zang Anniversary Volume. 1987 ISBN 90-247-3249-2
7. R. Negruţiu: *Elastic Analysis of Slab Structures.* 1987 ISBN 90-247-3367-7
8. J.R. Vinson: *The Behavior of Thin Walled Structures.* Beams, Plates, and Shells. 1988
   ISBN Hb 90-247-3663-3; Pb 90-247-3664-1

Mechanics

MECHANICS OF FLUIDS AND TRANSPORT PROCESSES
Editors: R.J. Moreau and G.Æ. Oravas

1. J. Happel and H. Brenner: *Low Reynolds Number Hydrodynamics.* With Special Applications to Particular Media. 1983 ISBN Hb 90-01-37115-9; Pb 90-247-2877-0
2. S. Zahorski: *Mechanics of Viscoelastic Fluids.* 1982 ISBN 90-247-2687-5
3. J.A. Sparenberg: *Elements of Hydrodynamics Propulsion.* 1984 ISBN 90-247-2871-1
4. B.K. Shivamoggi: *Theoretical Fluid Dynamics.* 1984 ISBN 90-247-2999-8
5. R. Timman, A.J. Hermans and G.C. Hsiao: *Water Waves and Ship Hydrodynamics.* An Introduction. 1985 ISBN 90-247-3218-2
6. M. Lesieur: *Turbulence in Fluids.* Stochastic and Numerical Modelling. 1987 ISBN 90-247-3470-3
7. L.A. Lliboutry: *Very Slow Flows of Solids.* Basics of Modeling in Geodynamics and Glaciology. 1987 ISBN 90-247-3482-7
8. B.K. Shivamoggi: *Introduction to Nonlinear Fluid-Plasma Waves.* 1988 ISBN 90-247-3662-5
9. V. Bojarevičs, Ya. Freibergs, E.I. Shilova and E.V. Shcherbinin: *Electrically Induced Vortical Flows.* 1989 ISBN 90-247-3712-5
10. J. Lielpeteris and R. Moreau (eds.): *Liquid Metal Magnetohydrodynamics.* 1989 ISBN 0-7923-0344-X

MECHANICS OF ELASTIC STABILITY
Editors: H. Leipholz and G.Æ. Oravas

1. H. Leipholz: *Theory of Elasticity.* 1974 ISBN 90-286-0193-7
2. L. Librescu: *Elastostatics and Kinetics of Aniosotropic and Heterogeneous Shell-type Structures.* 1975 ISBN 90-286-0035-3
3. C.L. Dym: *Stability Theory and Its Applications to Structural Mechanics.* 1974 ISBN 90-286-0094-9
4. K. Huseyin: *Nonlinear Theory of Elastic Stability.* 1975 ISBN 90-286-0344-1
5. H. Leipholz: *Direct Variational Methods and Eigenvalue Problems in Engineering.* 1977 ISBN 90-286-0106-6
6. K. Huseyin: *Vibrations and Stability of Multiple Parameter Systems.* 1978 ISBN 90-286-0136-8
7. H. Leipholz: *Stability of Elastic Systems.* 1980 ISBN 90-286-0050-7
8. V.V. Bolotin: *Random Vibrations of Elastic Systems.* 1984 ISBN 90-247-2981-5
9. D. Bushnell: *Computerized Buckling Analysis of Shells.* 1985 ISBN 90-247-3099-6
10. L.M. Kachanov: *Introduction to Continuum Damage Mechanics.* 1986 ISBN 90-247-3319-7
11. H.H.E. Leipholz and M. Abdel-Rohman: *Control of Structures.* 1986 ISBN 90-247-3321-9
12. H.E. Lindberg and A.L. Florence: *Dynamic Pulse Buckling.* Theory and Experiment. 1987 ISBN 90-247-3566-1
13. A. Gajewski and M. Zyczkowski: *Optimal Structural Design under Stability Constraints.* 1988 ISBN 90-247-3612-9

Mechanics

MECHANICS: ANALYSIS
Editors: V.J. Mizel and G.Æ. Oravas

1. M.A. Krasnoselskii, P.P. Zabreiko, E.I. Pustylnik and P.E. Sbolevskii: *Integral Operators in Spaces of Summable Functions.* 1976 ISBN 90-286-0294-1
2. V.V. Ivanov: *The Theory of Approximate Methods and Their Application to the Numerical Solution of Singular Integral Equations.* 1976 ISBN 90-286-0036-1
3. A. Kufner, O. John and S. Pučík: *Function Spaces.* 1977 ISBN 90-286-0015-9
4. S.G. Mikhlin: *Approximation on a Rectangular Grid.* With Application to Finite Element Methods and Other Problems. 1979 ISBN 90-286-0008-6
5. D.G.B. Edelen: *Isovector Methods for Equations of Balance.* With Programs for Computer Assistance in Operator Calculations and an Exposition of Practical Topics of the Exterior Calculus. 1980 ISBN 90-286-0420-0
6. R.S. Anderssen, F.R. de Hoog and M.A. Lukas (eds.): *The Application and Numerical Solution of Integral Equations.* 1980 ISBN 90-286-0450-2
7. R.Z. Has'minskiĭ: *Stochastic Stability of Differential Equations.* 1980 ISBN 90-286-0100-7
8. A.I. Vol'pert and S.I. Hudjaev: *Analysis in Classes of Discontinuous Functions and Equations of Mathematical Physics.* 1985 ISBN 90-247-3109-7
9. A. Georgescu: *Hydrodynamic Stability Theory.* 1985 ISBN 90-247-3120-8
10. W. Noll: *Finite-dimensional Spaces.* Algebra, Geometry and Analysis. Volume I. 1987 ISBN Hb 90-247-3581-5; Pb 90-247-3582-3

MECHANICS: COMPUTATIONAL MECHANICS
Editors: M. Stern and G.Æ. Oravas

1. T.A. Cruse: *Boundary Element Analysis in Computational Fracture Mechanics.* 1988 ISBN 90-247-3614-5

MECHANICS: GENESIS AND METHOD
Editor: G.Æ. Oravas

1. P.-M.-M. Duhem: *The Evolution of Mechanics.* 1980 ISBN 90-286-0688-2

MECHANICS OF CONTINUA
Editors: W.O. Williams and G.Æ. Oravas

1. C.-C. Wang and C. Truesdell: *Introduction to Rational Elasticity.* 1973 ISBN 90-01-93710-1
2. P.J. Chen: *Selected Topics in Wave Propagation.* 1976 ISBN 90-286-0515-0
3. P. Villaggio: *Qualitative Methods in Elasticity.* 1977 ISBN 90-286-0007-8